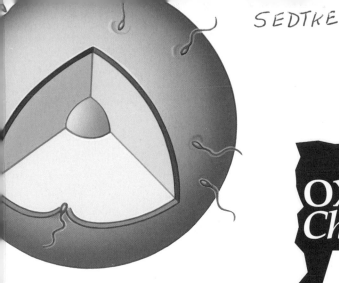

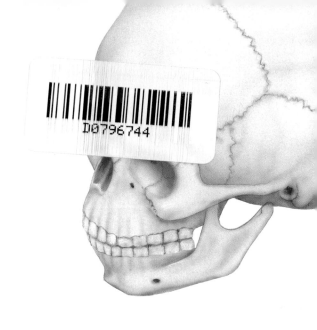

The OXFORD Children's A to Z of the

Human Body

Bridget and
Neil Ardley

OXFORD UNIVERSITY PRESS

Oxford University Press, Walton Street, Oxford, OX2 6DP

Oxford New York
Athens Auckland Bangkok Bombay
Calcutta Cape Town Dar es Salaam Delhi
Florence Hong Kong Istanbul Karachi
Kuala Lumpur Madras Madrid Melbourne
Mexico City Nairobi Paris Singapore
Taipei Tokyo Toronto

and associated companies in
Berlin Ibadan

Oxford is a trade mark of Oxford University Press

© Bridget and Neil Ardley 1996

First published in 1996
10 9 8 7 6 5 4 3 2 1

ISBN 0 19 910318 6 (hardback)
ISBN 0 19 910085 3 (paperback)

A CIP catalogue record for this book is available from
the British Library

Printed in Italy by G. Canale & C. S.p.A.
Borgaro T.se - TURIN

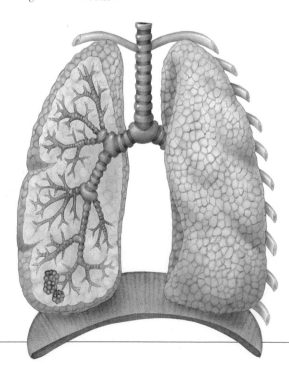

Acknowledgements

Picture research: Suzanne Williams

Abbreviations: t = top; b = bottom; l = left; r = right; c = centre;
back = background

Photographs

The publishers would like to thank the following for permission
to reproduce the following photographs:

Ace Photo Agency: 5r/Phototake, 10 heading panel/Phototake,
28-29/Dave Bunce;

Allsport: 50b/Ben Radford, 55b/Mike Powell;

Bruce Coleman: 45bl, 45br/Michael Freeman;

Collections: 9/Anthea Sieveking, 12/Anthea Sieveking;

Barry Etra: 60r (identical twins);

Sally & Richard Greenhill: 17, 62;

The Hutchison Library: 43/Sarah Errington;

Impact Photos: 51/David Reed;

Latha Menon: 24;

The Metropolitan Police: 22-23;

National Medical Slide Bank: 34tl;

Rex Features: 45tr/The Sun;

Science Photo Library: 6/CNRI, 7/John Durham, 10tl/Robert
Becker, 10tr/A.B.Dowsett, 10cl/NIBSC, 10cr/Dr Tony Brain, 13/Dr
Tony Brain, 14/Biophoto Associates, 15/Professors P.Motta &
S.Correr, 16tl/Bonnie Cosgrove, 16b/Adam Hart-Davis, 19/Gary
Parker, 23c/Claude Nuridsany & Marie Perennou, 25t/David
Scharf, 25, 26/J.C.Revy, 30t/Biology Media, 31/CNRI, 33/Robert
Becker/Custom Medical Stock Photo, 34br/BSIP VEM, 35, 36, 37t,
37bl, 37br/Astrid & Hanns-Frieder Michler, 38/Eric Grave, 39,
40-41b/David Scharf, 41t/David Scharf, 42/BSIP, 46/Martin
Dohrn, 47/Rory McClenaghan, 48/Jean Cox, 49/Scott
Carmazine, 50t/D.Phillips, 51ct/James Stevenson, 51cb/James
Stevenson, 52/CNRI, 54-55/Richard Wehr, 56/Professor P.Motta,
57/Chris Priest, 57/NASA, 58tr/Professor P.Motta/Dept. of
Anatomy, University 'La Sapienza', Rome, 58bl, 58tl/D.Phillips,
54-55/Richard Wehr, 60bl;

Science and Society Picture Library, **The Science Museum, London**: 4

Illustrations and diagrams

Michael Courtney: front cover cb, back cover c,
1tl and b, 2br, 3tr and bl, 9tl and b, 11c,
12c, 14tr, 16t, 18b, 20t, 21c, 23b, 26c
and b, 27b, 31c, 32c, 38c, 43b, 50,
38l, 61;

Clive Goodyer: front cover l r and t,
1c, 2c, 3cl and br, 4r and b, 5t, 6tr
and cr, 7l, 8t and b, 9tr, 11br,
12br, 13b and r, 14l, 15l and heading panel, 17l, 18c, 19l, 20b, 21b, 22t
and c, 25bl and tr, 27t, 30b, 31back and
br, 34/35b, 36l, 37l and r and heading
panel, 38t, 42b, 42/43t, 44, 46, 48, 49cl, 52tl, 54cl and b, 55tl
and br, 56tl and br, 59t, 60tl, 62l, 63b, 64l;

John Haslam: 9tl;

Steve Weston: back cover tl and r, 1tr, 2bl, 12c, 14l, 15br, 24r,
28/29, 32tr, 33t and b, 36c, 39b, 40c, 41t and b, 47, 51, 52ct, 53,
59b

Dear Reader

What will beat about 100 thousand times today, and maybe 10 million times during your lifetime? It's your heart. But do you know what your heart is, and why it keeps on beating so much?

Your heart is part of your body, and it's very important to know all about your body. Its many parts work together to keep you alive so that you can do and enjoy so many things – like eating your favourite foods and playing games, as well as seeing and hearing the world around you.

Use this book to discover the parts of your body, where each part is, and how it works. Read about how your body grows and keeps healthy. And learn about the names of illnesses that people can get, why an illness happens and how it can be cured.

We hope that this book will help you to understand your body and to look after it, so that it will last a very long time!

Bridget and Neil Ardley

abdomen

Your abdomen is the part of your body below your chest. It is also called your belly. Your stomach, intestine, liver, kidneys, pancreas and gall bladder are all in your abdomen.
See also **human body**.

abrasion

Abrasion is another name for **graze**.

abscess

An abscess is a painful swelling that can develop anywhere in or on the body. It forms when **bacteria** infect a place, which swells and fills with **pus**. If the abscess is inside the body a doctor may treat it with **antibiotics**.

A Chinese acupuncture statue. Acupuncture started in China about 5000 years ago and is still used today.

acne

Acne is a collection of **spots** and **blackheads** that usually occur on the face, chest, back or shoulders. It happens when oil **glands** in the **skin** block up. Acne is common among teenagers. It can be treated with special lotions, creams and soaps, but it often gets better as you grow up. A doctor will treat acne if it is very troublesome.

A section of skin showing a blackhead ①, and a spot ②. These happen when oil glands and sweat glands become blocked.

A modern acupuncture needle.

acupuncture

Acupuncture is a treatment in which needles are stuck into different parts of the body. The needles go only a little way into the skin. Acupuncture often relieves pain, and many people believe that it cures illness.

Adam's apple

The Adam's apple is a small lump at the front of your neck. It is the front of your **larynx**, or voice box. Men usually have a bigger Adam's apple than women.

adenoids

Your adenoids are small lumps at the back of your nose. They help your **tonsils** to protect your **lungs** from **infection**. Sometimes the adenoids become infected and swollen and make it difficult to breathe through the nose. People with swollen adenoids may sound as if they have a cold.

adolescence

Adolescence is the time during which a child develops into an adult. During this time children are called adolescents.
See also **puberty**.

sweat pore

sweat gland

① ②

hair

oil gland

adrenaline

Adrenaline is a substance that your body makes to help you when you are frightened, angry or excited. It is a **hormone** produced by certain **glands**. Adrenaline speeds up your breathing and makes your heart beat faster and your muscles work harder. This may help you to run faster to escape danger or to win a race.

ageing

Ageing means growing old. Bones may break more easily when people get old. Muscles get smaller and the **joints** may get stiff, making it hard to walk. The hair often turns grey or white, and the skin usually starts to wrinkle. Many old people often cannot see, hear or remember things as well as they did when they were younger.

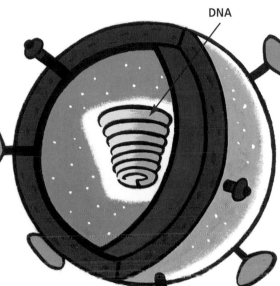

This is a diagram of the AIDS virus, magnified many times.

DNA

outer shell of virus

AIDS

AIDS is a very serious disease that stops the body fighting other diseases. It is caused by a **virus** known as **HIV**, but even if someone has the virus, they may not develop AIDS. The AIDS virus can spread by passing from an infected person to someone else. This may happen through **sex**, an infected blood transfusion or a needle already used by someone with the AIDS virus. A woman with AIDS or the AIDS virus may pass it to her unborn baby. At the moment there is no cure for AIDS, but scientists are working hard to find one. AIDS stands for Acquired Immune Deficiency Syndrome.

alcohol

Alcohol is a liquid in drinks like beer and wine, and in spirits like whisky. Drinking too much stops the body working properly, so people should not drink if they are going to drive. People sometimes behave badly when they have drunk too much alcohol. Heavy drinking can also damage parts of the body, like the **liver**.

allergy

You have an allergy if certain things make you sneeze or make you get an itchy **rash** on your skin. Many different things can cause allergies. Some people are allergic to particular foods and others to animal fur. Hay fever is an allergy caused by pollen from grasses or flowers. If you know you are allergic to something in particular, it is best to try to avoid it if you can. Doctors sometimes give people tablets or **injections** to treat allergies.

Many old people remain very active and take part in sporting events, such as a marathon.

This is a photograph of grains of pollen magnified many times. Pollen grains cause an allergy in some people.

ambidextrous

If you are ambidextrous then you are able to use each of your hands with the same skill.

amnesia

Amnesia is loss of **memory** that can be caused by a serious shock or an accident.

anaemia

Anaemia is an illness that makes a person look pale and feel tired. It happens if the blood does not carry enough **oxygen** around the body. Your body needs oxygen to combine with your food and make energy, which keeps your body going. There are several kinds of anaemia which doctors treat in different ways.

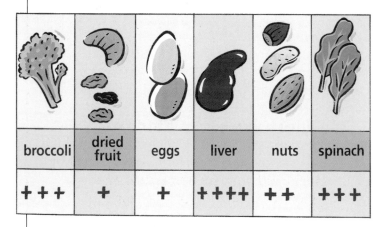

broccoli	dried fruit	eggs	liver	nuts	spinach
+++	+	+	++++	++	+++

Anaemia can be caused by not having enough iron in your diet. The table shows the level of iron in certain foods (the more + signs there are, the more iron). You must have iron in your diet to help your blood carry oxygen around your body.

anaesthetic

An anaesthetic is a drug used in **operations** and other forms of treatment, so that you do not feel any pain. A dentist may give you an anaesthetic by injection. An anaesthetic that removes feeling from the part of the body to be treated is called a local anaesthetic. An anaesthetic that makes a person unconscious during an operation is called a general anaesthetic.
See also **unconsciousness.**

A bacterium before and after treatment with antibiotics, magnified many times.

ankle

Your ankle is the part of your body between your foot and your leg.
See also **human body.**

anorexia nervosa

Anorexia nervosa is an illness which makes a person stop wanting to eat. Someone suffering from it may become very thin. They have to be treated by a doctor to help them to learn to eat again.

antibiotic

An antibiotic is a drug which attacks **bacteria.** The doctor may give you antibiotics if you have an illness caused by bacteria, such as **bronchitis. Penicillin** is a common antibiotic.

antibodies

Antibodies are substances produced by your body that kill harmful **germs.**
See also **vaccination.**

antiseptic

An antiseptic is a substance which kills many **germs** that cause disease. Antiseptics are used to clean both wounds and doctors' instruments.

anus

Your anus is the opening where waste material leaves your body as **faeces** when you go to the toilet.

to head
from heart
to body

The aorta is shaped like an arch as it leaves the heart.

aorta

The aorta is the largest **artery** in the body. Blood flows from the **heart** through the aorta, which then branches to carry blood to all parts of the body.

The arteries in the body vary in width.

small artery

large artery

artery

appendicitis

Appendicitis is an illness that is caused by **inflammation** of the appendix.

appendix

Your appendix is a small tube growing out of your large **intestine**. Millions of years ago the appendix might have been used in **digestion** but it seems to have no use now. Sometimes it becomes sore and inflamed, causing appendicitis. This illness can be dangerous, and the appendix may have to be removed by an **operation**.

arm

Your arm is made up of the upper arm, the elbow, the forearm and the wrist. It is joined to the main part of your body at the shoulder and to your hand by the wrist joint. See also **human body**.

artery

An artery is a tube which carries **blood** away from your heart. You have many arteries, which take blood to all parts of your body. See also **circulation**.

arthritis

Arthritis is a disease that affects the **joints** of the body making them stiff and painful.

aspirin

Aspirin is a medicine that you can take to relieve a headache or a pain somewhere else in your body. It is usually taken in the form of tablets.

asthma

Asthma makes breathing difficult. When someone has an asthma attack, they begin to wheeze and cough. Asthma may be caused by an **allergy** or by an **infection**. It is often treated by breathing in a special spray.

If you suffer from asthma, you can use drugs from an inhaler to help you breathe.

athlete's foot

Athlete's foot is flaky, itchy patches of **skin** between the toes. It is caused by an **infection** from a **fungus**, and is made worse when the feet get hot and sweaty. Athlete's foot can be cured by special powders and creams which kill the fungus.

B

baby

A baby is a very young child.
See also **birth**.

backache

Backache is pain in your back. It can be caused by standing or sitting incorrectly. Sometimes too much exercise can give you backache. Damage to the **spine** also causes backache.

See also **posture**.

backbone

Backbone is another name for **spine**.

bacteria

Bacteria are living things that are so small you cannot see them. A single one is called a bacterium. Bacteria live all around you, for example in air and water, and they also live inside you. Most bacteria are harmless but some cause diseases if they enter your body. This is why it is important to clean a cut in your skin. **Tuberculosis** is an example of a disease caused by bacteria. One way your body can fight harmful bacteria is by making **antibodies**, or the doctor may give you **antibiotics**.
See also **germ**.

bad breath

Sometimes a person's breath has an unpleasant smell. This can happen if the teeth are dirty and have bits of food stuck between them. Strong tasting food can also make breath smell. If you clean your teeth often and carefully, it helps to prevent bad breath. Halitosis is another name for bad breath.

Bacteria have many different shapes as these enlarged drawings show.

balance

Balance is your ability to stay upright when you stand or walk, even with your eyes closed. It stops you falling over. Inside your ears there are tubes filled with liquid. The liquid, which moves around as you turn or tilt your head, presses against **nerves** which send messages to your **brain** so that you always know which is the right way up. Your brain also uses messages from your eyes, muscles, joints and toes to help you keep your balance.
See also **travel sickness**.

baldness

Baldness occurs if the hair on a person's head falls out. It may happen as a result of illness, after which the hair will often grow back again, or it may happen during old age, when the hair usually does not grow back.
See also **hair**.

belch

A belch is a noise caused by air coming up from your **stomach** and out through your mouth. It is also called a burp.

belly

Belly is another name for **abdomen**.

biceps

The biceps is the **muscle** in your upper arm that makes your arm bend.

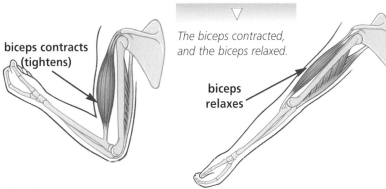

biceps contracts (tightens)

biceps relaxes

The biceps contracted, and the biceps relaxed.

bile

Bile is a greenish liquid which helps your **liver** to digest **fat**. It is stored in your **gall bladder**. If you are sick, some bile may come up. It has an unpleasant bitter taste.
See also **digestion**.

birth

A mother gives birth to a baby after it has been inside her **womb** for about nine months. By this time the baby has grown enough to live separately. Muscles around the womb begin to push the baby out, usually head first, through the **vagina**. This is called labour, and it may take several hours. When the baby is born, it is still joined to its mother by the **umbilical cord**. Soon after, this cord is cut and tied. The remains of it form your **navel**, or tummy button.
See also **pregnancy**.

blackhead

A blackhead is a small black **spot** on your skin. Keeping your skin very clean and eating healthy food may help to stop blackheads forming.
See also **acne**.

bladder

Your bladder is a bag inside you that stores liquid made up of water your body does not need and waste substances from your blood. This liquid, called urine, comes from your **kidneys**. You can feel when your bladder is full of urine and you need to go to the toilet.
See also **human body**.

bleeding

Bleeding happens when blood leaks from a cut or **wound** in your **skin**. Some diseases or injuries can cause bleeding inside the body. This is called internal bleeding and is treated by a doctor.
See also **blood, period, scab**.

blindness

Blindness, or visual impairment, is the loss or lack of sight in one or both eyes. A few people are born blind. Some become blind because their eyes are damaged. Certain diseases may cause blindness, and old people may go blind. Sometimes blindness can be cured.

womb

placenta

umbilical cord

vagina

A baby is usually born head first.

spine

Blood

white blood cell

Blood is the red liquid inside your body. It flows through tubes called blood vessels, and is pumped all around your body by your **heart**. Blood carries substances, which come from the food you eat, and oxygen, which comes from the air you breathe, to make you stay alive and grow. It also helps to take waste products away. Blood is made up of blood cells, or corpuscles, and small particles called platelets, which all float in a yellowish liquid called **plasma**.

There are two kinds of blood cells – red and white. Red cells contain haemoglobin, which gives blood its red colour and also carries oxygen. White cells mainly fight infection. The platelets help to heal wounds.

platelet

red blood cells

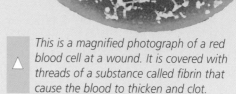

This is a magnified photograph of a red blood cell at a wound. It is covered with threads of a substance called fibrin that cause the blood to thicken and clot.

A blood clot forms when blood thickens. This happens when you cut or **wound** yourself and blood leaks onto your skin. The clot forms to stop your body from losing too much blood. A clot may also form in the blood vessels inside the body. If the blood vessels have become hard and narrow, a clot may block a blood vessel and this can cause a heart attack.

Blood pressure is a measure of how much the blood presses on the walls of your **arteries** as it is pumped around your body. This pressure usually increases when your heart beats faster.

See also **circulation**, **heart**, **wound**.

Everybody's blood belongs to one of four main blood groups, depending on certain substances in the red cells. Different types of blood, or blood groups, are called A, B, AB and O. If someone is given extra blood by a blood transfusion, the blood must be of the same or a matching blood group as their own.

The table shows matching blood in the ABO system of blood groups.

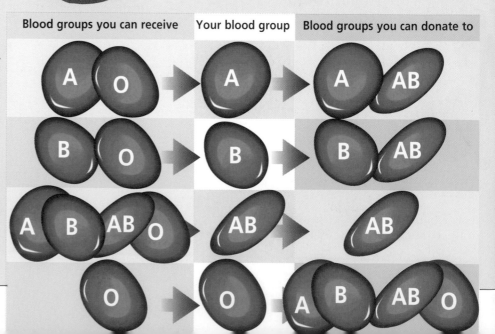

Blood groups you can receive	Your blood group	Blood groups you can donate to
A O	A	A AB
B O	B	B AB
A B AB O	AB	AB
O	O	A B AB O

10

femur →

blushing

Blushing is when your face suddenly becomes red and hot because you are angry or embarrassed. It is caused by blood rushing near to the surface of your skin.

body odour or b.o.

Body odour is the unpleasant smell of a person's body when it has been sweating and the sweat has grown stale. The smell is caused by **bacteria**.
See also **sweat**.

boil

A boil is a small **abscess** on the skin. Boils often heal on their own and they should not be squeezed. Sometimes boils are treated with **antibiotics**, or a hot pad may be put on them to soften the skin and relieve pain.

bone

Bones are the hard parts of your body. They form your **skeleton**, which is the framework that supports your body. Bones are also important because they store **minerals** that your body needs, and inside bones a material called marrow makes part of your blood.

Although your bones are hard, they contain living cells. Bones grow, and if they break or **fracture**, the broken ends slowly grow back together.

An arm or leg with a broken bone is often wrapped in plaster to make sure that the bone grows straight as it mends.

hollow shaft containing marrow

compact bone

spongy bone

cartilage

stirrup bone (of the ear)

The smallest bone in the body is in the ear and the largest is the thigh bone, or femur.

bowel

The bowel, or bowels, is another name for the large **intestine**.

brace

A brace is a special framework which a dentist or orthodontist may fix to your teeth to help them to grow straight and even.

brain

Your brain is in your head and controls the whole of your body and nearly everything you do. It contains many millions of **cells**.

Each section of your brain looks after a different part of the body. The largest section, called the cerebrum, controls the senses, intelligence, movement and feelings. A part called the cerebellum controls your **muscles** and helps with **balance**.

Nerves all over your body send messages to the brain. The brain then sends signals out along more nerves to all the different parts of the body to make them work. Your brain sorts and stores information, so that you can think, learn and remember. Your brain works all the time, even when you are asleep, so that you keep breathing and your **heart** continues to beat.

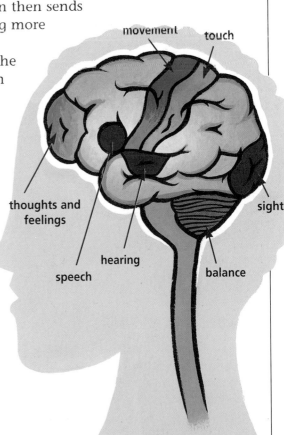

movement touch
thoughts and feelings
sight
hearing speech balance

Different areas of the brain control different senses and your actions.

breast

The breasts are two **glands** on the front of a woman's chest. When a baby is born, the breasts make milk for it to feed on. The baby sucks the milk through the **nipple** at the tip of the breast. Girls start to grow breasts during **adolescence** and **puberty**.

◁ *A mother breast-feeding her baby.*

breathing

Breathing is drawing air into your **lungs** and letting air out of your lungs through your nose and sometimes through your mouth. Your **diaphragm** and **rib** muscles move the walls of your lungs to suck in and let out air.

Your body needs **oxygen** to produce energy. Oxygen from the air goes into your lungs and then into your blood. Inside your body energy is produced from the food you eat and the oxygen you breathe. Some of the oxygen is changed into a gas called **carbon dioxide**. This returns to your lungs and leaves your body when you breathe out.
See also **circulation**, **respiration**.

bronchitis

Bronchitis is an illness that makes you cough a lot and makes breathing difficult. It is caused by **germs**, and can also be caused by **smoking**. These produce soreness of the bronchi, which are the air passages inside your **lungs**. Bronchitis is sometimes treated with **antibiotics**.

▽ *Breathing in and breathing out.*

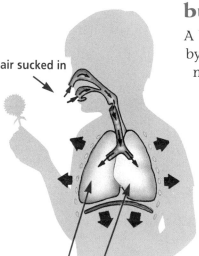

air sucked in

lungs get bigger

carbon dioxide breathed out

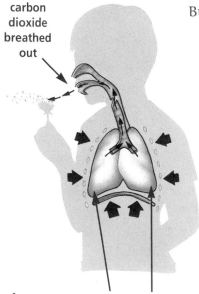

lungs get smaller

bruise

A bruise is a dark mark on your skin, which may feel painful. It is caused by a hard knock which breaks some of the blood vessels under your skin. **Blood** leaks out of the blood vessels and causes the skin to change colour. Bruises usually disappear quite quickly.

burn

A burn is damage to your skin caused by heat or by certain chemicals. A mild burn usually heals without leaving a mark, but more serious burns cause blisters which are swellings filled with blood or **serum**. Some deep burns may need to be treated by a doctor. A burn caused by hot water or steam is called a scald. If you burn or scald yourself, ask an adult for treatment right away.

burp

Burp is another name for **belch**.

buttocks

The buttocks are the rounded parts of your bottom that you sit on. They contain the big **muscles** that help you to stand up and sit down.
See also **human body**.

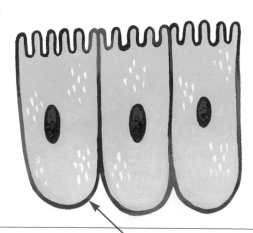

stomach cell

C

calf

Your calf is the back part of your lower **leg**, between the knee and the ankle.
See also **human body**.

cancer

Cancer is a serious disease in which the **cells** in a part of the body begin to grow in an unusual way. They make swellings called **tumours** which harm the body. Sometimes cells may break off the tumour and start growing as another tumour, so that the cancer spreads. Cancer can attack almost any part of the body.

There are many different types of cancer. **Leukaemia** is a disease of the **blood** caused by cancer of the **bone** marrow. It is known that certain things, such as **smoking** and **radiation**, can cause cancer. Sometimes doctors can cure cancer by removing the tumour in an operation. **Drugs** and **X-rays** are also used to treat cancer. You cannot catch cancer from other people.

canine

Your canines are pointed **teeth** used for biting.

capillary

A capillary is the smallest kind of blood vessel.
See also **blood**, **circulation**.

cancer cells

normal cells

You can see how different these cancer cells, green in this picture, look from normal cells.

carbohydrate

Carbohydrate is a substance in **food** that gives you energy. Foods that contain carbohydrate include bread, potatoes, rice, cereals and sugar.

carbon dioxide

Carbon dioxide is a gas produced by your body when you use food and drink. It leaves your body when you breathe out.
See also **breathing**.

cartilage

Cartilage is a smooth, white, elastic material that covers the ends of the bones in **joints** so that they move easily.

cell

Cells are the smallest living units which make up your body. Each part of your body consists of many millions of cells.

There are several different types of cells, including bone cells, skin cells, brain cells, nerve cells and blood cells. The same cells grouped together are called **tissue**.

muscle cell

nerve cell

The different kinds of cells in your body have different shapes.

blood cell

bone cell

central nervous system

Your central nervous system is made up of your **brain** and **spinal cord**.
See also **nerve**.

cerebral palsy

Cerebral palsy is a serious illness in which a person's **muscles** do not work properly. It is caused by damage to the part of the **brain** that controls movement. This damage happens before, during, or very soon after the person's birth. You cannot catch cerebral palsy from another person. Sometimes treatment can help people with cerebral palsy.

brain

spinal cord

cervix

The cervix is a part of the **womb**.

chest

Your chest is the upper part of your body between your neck and your **abdomen**. Your **heart** and **lungs** are inside your chest, enclosed by your **ribs**. The chest is also called the thorax.
See also **human body**.

chicken pox

Chicken pox is a disease caused by a **virus**. Many children get chicken pox. You catch it from other people by breathing in the virus. It causes a **rash**.

chilblain

A chilblain is an itchy swelling of the skin caused by cold. People can get chilblains on their fingers, toes or ears.

chill

Chill is a word that some people use to describe a shivery feeling. It is not an illness.

> Your epiglottis is the flap that stops food going down your windpipe.

food enters windpipe

choking

Choking happens when food you are swallowing goes down into your **windpipe** instead of the **gullet** to your **stomach**. It can make you cough and become short of breath.
See also **epiglottis**.

epiglottis open

windpipe

gullet

cholera

Cholera is a serious disease caused by drinking water which is not clean. It mainly occurs in poor countries.

cholesterol

Cholesterol is a fatty substance made by the **liver**. It helps your body to digest food. But too much cholesterol in the blood stops the blood flowing properly and could cause a heart attack. People with heart problems often eat special food to try and reduce the amount of cholesterol in their blood.
See also **heart**.

chromosome

There are tiny chromosomes in nearly every cell in your body. They are made up of DNA and carry the genes that decide what you look like. You inherit your chromosomes from your father and mother, which is why you may look like your parents.
See also **DNA, gene, sex**.

Chromosomes magnified many times.

Circulation

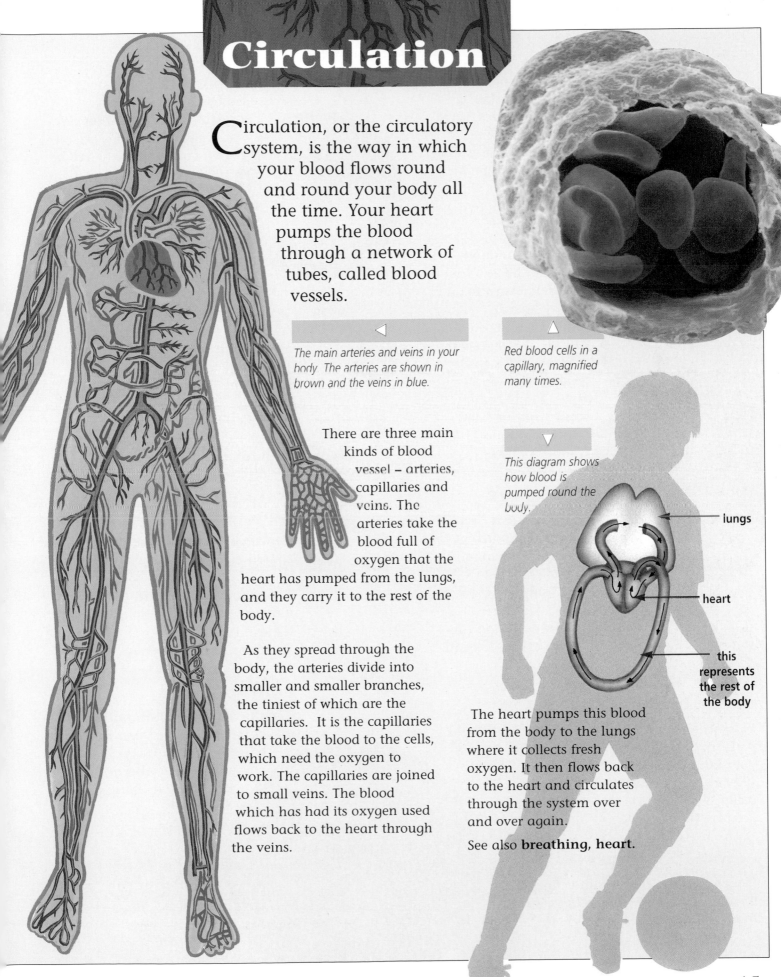

Circulation, or the circulatory system, is the way in which your blood flows round and round your body all the time. Your heart pumps the blood through a network of tubes, called blood vessels.

◁

The main arteries and veins in your body. The arteries are shown in brown and the veins in blue.

△

Red blood cells in a capillary, magnified many times.

There are three main kinds of blood vessel – arteries, capillaries and veins. The arteries take the blood full of oxygen that the heart has pumped from the lungs, and they carry it to the rest of the body.

As they spread through the body, the arteries divide into smaller and smaller branches, the tiniest of which are the capillaries. It is the capillaries that take the blood to the cells, which need the oxygen to work. The capillaries are joined to small veins. The blood which has had its oxygen used flows back to the heart through the veins.

▽

This diagram shows how blood is pumped round the body.

lungs

heart

this represents the rest of the body

The heart pumps this blood from the body to the lungs where it collects fresh oxygen. It then flows back to the heart and circulates through the system over and over again.

See also **breathing, heart.**

15

circumcision

Circumcision is a simple operation sometimes carried out on a boy's **penis**. The foreskin, a fold of skin covering the end of the penis, is cut away. This may be done because the foreskin is too tight, or it may done for religious reasons.

cold, or common cold

When you have a cold, or common cold, you feel shivery, your nose runs, you sneeze and you may have a headache and sore throat. It is caused by a **virus** and is easily caught from other people. A cold usually only lasts for a few days and is not a serious illness.

This photograph shows some cold viruses magnified many times.

coldsore

A coldsore is a small reddened patch of skin near the mouth or nose. Sometimes blisters form as well. You may get a coldsore when you have a cold. It is not serious and can be treated with cream from the doctor.

collar bone

Your collar bone is the bone that reaches from your shoulder to your neck across the front of your body. You can easily feel it on each side below your neck. Another name for collar bone is clavicle.
See also **skeleton.**

colour blindness

People who are colour blind cannot tell the difference between certain colours, most often red and green.

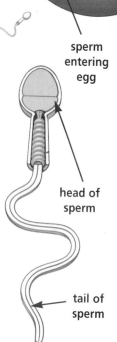

centre of egg

egg

sperm entering egg

head of sperm

tail of sperm

This diagram shows a single sperm entering an egg at conception.

coma

A coma is like a very deep sleep. It is caused by illness or by an accident, and can last for days or even months. If a person is in a coma, they have to be looked after in hospital.

conception

Conception is the moment when a male **sperm** joins with a female **egg** inside a woman's body and a baby starts to grow. Sometimes the sperm and the egg are put together by doctors in a laboratory. Babies conceived by this method are called **test-tube babies**. Another name for conception is fertilization.
See also **reproduction.**

concussion

Concussion is a shaking of the **brain** caused by a hard bang to the head. It may cause a person to go to sleep, or become unconscious. A doctor. may treat someone who is thought to have concussion.
See also **unconsciousness.**

conjunctivitis

Conjunctivitis is **inflammation** of the white part of your eye, which becomes red and sore. It can be caused by an **allergy** or by **infection**. It is also called pink eye.

If you can see the number 57, you have full-colour vision. If you see 35, you cannot tell the difference between red and green.

contraception

Contraception means stopping male **sperm** joining with a female **egg** so that a baby does not start to grow. People use contraception when they do not want to have a baby.
See also **conception, reproduction.**

cough

You cough when you suddenly push air out of your lungs. You do it automatically, without thinking about it, when you need to clear the tubes through which you breathe. Some illnesses, like **bronchitis**, cause coughing.

cramp

Cramp is a pain in a muscle. It can happen when you are taking exercise. Rubbing the muscle often helps to get rid of cramp.
See also **muscle.**

Deaf people can understand each other very easily using sign language.

cut

A cut is a kind of **wound**. A cut should be cleaned and covered to stop dirt getting into it.

cystic fibrosis

Cystic fibrosis is a serious disease that some people are born with. It affects the lungs and the digestive system. You cannot catch cystic fibrosis from other people. It is an inherited disease, which means that it is passed on to a person from their parents.

D

dandruff

Dandruff is a collection of small white flakes in your hair. The flakes are dead skin from the top of your head. Special shampoos can help to get rid of dandruff.

deafness

Deafness is not being able to hear properly in one or both ears. It can be caused by damage to the inside of the ear or by disease. Some people may be born deaf and others become deaf as they grow older. People who are hard of hearing may be helped with a hearing aid.
See also **dumbness, ear.**

death

Death is when life ends. When a person grows old, parts of the body stop working properly and the person dies. Death may also happen as the result of an accident or an illness.

dermatitis

Dermatitis is **inflammation** of the skin and shows as red, itchy patches. It may be caused by an **allergy** and can be treated with special creams.

diabetes

Diabetes is an illness in which a person's body does not produce the right quantity of a **hormone** called insulin. Insulin is needed to control the amount of sugar in your blood. People with diabetes may have to inject themselves with insulin every day.
See also **pancreas.**

diagnosis

When your doctor decides what is the matter with you, they are making a diagnosis.

dialysis

Dialysis is a treatment that removes waste materials from the blood. The **kidneys** normally do this, but if a person's kidneys are not working properly they might have to use a dialysis machine.

diaphragm

Your diaphragm is a large flat muscle between your **chest** and your **abdomen**. It helps to control your **breathing**.

diarrhoea

When you to go to the toilet and pass very watery waste, you have diarrhoea. It is usually caused by an **infection** in the **intestine**. A doctor will treat diarrhoea if it does not clear up quickly. See also **digestion**, **faeces**.

diet

Your diet is the mixture of food you eat. A healthy diet is food that is good for your body. But if you are 'on a diet', it means that you have to eat special food, for example to try and lose weight. See also **food**.

digestion

Digestion is the way in which your body changes food and drink into substances which can enter your blood.

From your mouth, food and drink enter your digestive system and pass slowly through your body. As you digest your food and drink, substances from them enter your blood, which carries the substances to all parts of your body. The substances give you energy, which makes all your body parts work.

Waste material in your food and drink is not used; it passes right through your digestive system and leaves your body.

If you eat too much or too fast, your digestive system may not work properly and you will get a pain called **indigestion**.

diphtheria

Diphtheria is a very serious disease of the throat. Nowadays it is rare because most children are given a **vaccination** against it.

disability

A disability means that part of the body does not work properly. People may suffer from mental disability or physical disability.

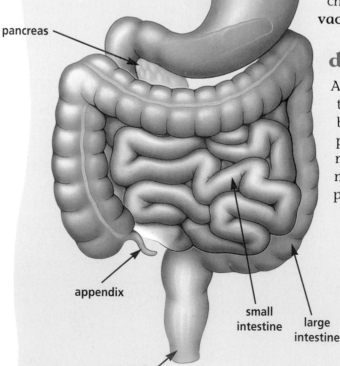

mouth
gullet
stomach
pancreas
appendix
small intestine
large intestine
rectum

The entire adult digestive system is about nine metres long.

disease

If you have a disease, it means that some part of your body is not working properly. Some diseases are caused from infection by **bacteria**, or by a **virus**. People are sometimes born with a disease which has been passed on to them by one or both of their parents. Diseases can be very mild, like the common cold, or very serious, like cancer.

DNA

DNA is a short way of saying deoxyribonucleic acid, which is a natural chemical found in nearly every cell in your body. Your **chromosomes** are made up of DNA. The DNA carries your **genes** which control what your cells do and the way in which you grow.

◁ *When highly magnified, DNA looks like a twisted ladder.*

Down's syndrome

Down's syndrome describes a disability that some people are born with. It makes them look different from other people, and they take longer to learn how to do things. It happens because a person is born with an extra **chromosome**.

▷

A Down's syndrome child playing with his mother and a friend.

drugs

Drugs are chemicals that people can take to cure certain diseases, and to stop pain. They should only be taken if a doctor says so.

There are some drugs that it is against the law to take. These drugs can cause illness or even death. A person who takes these kinds of drugs may begin to feel that they must keep on taking them: this is called drug addiction. Someone who becomes addicted to any drugs needs special treatment to help them to stop taking the drugs.

dumbness

Dumbness is a word that is sometimes used to describe people who have speech difficulties. People who are born hard of hearing often have problems in speaking. Special teachers help people with speech difficulties to learn to speak.

dyslexia

People who have dyslexia have difficulty in reading and writing because their brain muddles up letters and words. Being dyslexic does not mean that a person is not clever, and dyslexic people can often be helped by special teaching.

◁

Sign language, or finger spelling, is used by people who are hard of hearing or who have speech difficulties. These signs spell HELLO.

19

E

ear

Your ears let you hear sounds. Sound travels as waves through the air and enters your ears. The sound waves cause parts in your ears to send messages to your **brain** and you hear. Parts of your ears also help you to **balance**.

eczema

Eczema appears as red, itchy and sore patches on the **skin**. It may be caused by an **allergy** or it may be inherited (passed on at birth by a person's parents).

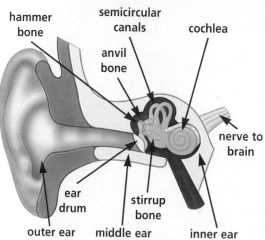

As well as the outer part you can see, your ear has a middle and inner part.

hammer bone
semicircular canals
anvil bone
cochlea
nerve to brain
ear drum
stirrup bone
outer ear
middle ear
inner ear

egg or egg cell

An egg is the cell in a woman that combines with the sperm from a man at **conception**. It is also called an ovum. See also **ovary**, **reproduction**.

elbow

Your elbow is the joint in the middle of your **arm**. See also **human body**.

embryo

An embryo is an unborn baby during the first eight weeks after **conception**. After eight weeks, the embryo is called a **foetus**.

enzyme

Enzymes are substances in your body which help the processes that make your body work. For example, there are enzymes in your **stomach** that help you to digest your food.

This table shows some of the enzymes produced by the body.

ENZYME	PRODUCED BY	ACTION
amylase	mouth	breaks down starches in food
pepsin	stomach	breaks down proteins in food
trypsin	pancreas	breaks down proteins
lipase	intestine	breaks down fats

epidemic

When a lot of people catch the same disease, such as chicken pox or influenza, at the same time, it is called an epidemic.

epiglottis

Your epiglottis is a flap which covers the top of your windpipe. It opens when you breathe, and closes when you swallow food. This is to stop the food going down the wrong way. See also **choking**.

epilepsy

Epilepsy is an illness of the **brain**. People with epilepsy sometimes have fits, which may cause them to become unconscious and to twitch and shake because their muscles stiffen. Someone who suffers from epilepsy has to be treated by a doctor. See also **unconsciousness**.

exercise

Exercise is moving your body about in different ways, such as running, dancing and swimming. Regular exercise keeps you fit, which means that your body is healthy and working properly.

There are two kinds of fats – saturated and unsaturated. Eating too much saturated fat may increase the risk of heart disease.

eye

Your eyes let you see things. Light rays come from an object and enter each eye through the pupil. The light is focused on to the back of the eye by the lens, where a picture of the object is formed. This part of your eye, called the **retina**, then sends messages to your **brain** and you see the object.

Your two eyes send a slightly different view of the object to your brain. Your brain combines the two views and this enables you to judge how far away the object is. Your eyelashes and eyelids help to keep dust out of your eyes.

The lens in each eye forms an upside-down picture on the retina. Your brain turns it the right way up.

eyeball
pupil
nerve to brain
iris
lens
retina

Milk, cheese, butter, chocolate and beef contain saturated fat.

Sunflower seeds, sweetcorn, some kinds of nuts and fish contain unsaturated fat.

faeces

After your body has taken everything it needs from your food, what is left forms into soft, brown faeces and leaves your body when you go to the toilet.

fainting

When a person faints they fall down and become unconscious for a short time, as if they were asleep. Fainting happens because the heart is not able to pump enough blood to the brain for some reason. When the person falls to the ground, the blood rushes back to their head and they soon wake up. 'Passing out' is another word for fainting. See also **unconsciousness**.

fat

Fat is a substance in food that gives you energy and helps you to grow. Dairy products like butter, milk and cheese contain fat. Eating too much fat can make your body get too big or fat.

fatigue

Fatigue is another word for tiredness.

femur

Your femur is the bone in your thigh. It is the longest bone in your body. See also **skeleton**.

fertilization

Fertilization is another word for **conception**.

fever

You have a fever when your body gets hotter than it should and you have a high **temperature**. Having a fever often means that you have an **infection**.

fibre

Fibre is the part of your food that you cannot digest. It comes from fruit and vegetables. You need to eat fibre because it helps your body to digest all the other kinds of food. Fibre is also called roughage.
See also **digestion**, **food**.

finger

Your fingers are joined to your hands. You have four fingers and a thumb on each hand.
See also **human body**.

fingerprint

Your fingerprints are the patterns of tiny lines on the tips of each of your fingers and thumbs. No two people have the same fingerprints, not even **twins**.

first aid

If someone becomes ill or has an accident, first aid is help that they can be given before a doctor or ambulance arrives. It is very useful to learn how to give first aid because it might help to save someone's life.

fit

If a person has a fit, they may twitch and shake as their muscles stiffen and they may become unconscious. **Epilepsy** can cause fits.
See also **unconsciousness**.

flu

Flu is short for **influenza**.

unconsciousness
or death

°C

fever
(37.7°)

normal
(about 36.5°)

40°

35°

30°

death

25°

There are special thermometers made to measure body temperature.

foetus

A foetus is a baby that has been in the womb for more than eight weeks. See also **embryo**.

follicle

A follicle is a little pit in your **skin** from which a single **hair** grows.

food

Food is what you eat to keep your body living and growing. There are several main substances in food that you need so you can keep healthy.

Your body needs **protein** to make you grow. Foods that contain protein include fish, meat, milk, beans and eggs. You also need to eat **carbohydrates** to give you energy, and you can get these from bread, cereals, sugar and potatoes. **Fat** is another important food which also gives you energy and keeps you warm. Fat comes from butter, milk and cheese (and many other foods).

Fibre, or roughage, comes from plants and cannot be digested, but you need it to help the other kinds of food through your digestive system. Plenty of vegetables and fruit will give you the fibre you need. Your body also needs small amounts of **minerals** and **vitamins** to stay healthy, grow, and repair itself if anything goes wrong. These are found in vegetables and fruits as well as in eggs, milk and fish. If food is not kept clean, or is not fresh or not cooked properly, it can give people **food poisoning** and make them ill. This is because **bacteria** have got into the food and so into the body of the person eating it.

These are the four main substances in food.

PROTEIN

food poisoning

Food poisoning is an illness that may cause **vomiting** and **diarrhoea**. Food poisoning is usually caused by eating food that is not clean, not fresh or not cooked properly.

CARBOHYDRATE | FAT | FIBRE

freckle

A freckle is a brown spot on the surface of the **skin**.

frostbite

Frostbite is damage caused to parts of the body by extremely cold weather. The parts affected lose all feeling, go very white and may later blacken. Frostbitten skin must be warmed very gently and covered with clean bandages as it easily becomes infected.

fungus

A fungus is a tiny plant that may grow on parts of the body and cause diseases such as **athlete's foot**. Mushrooms and toadstools are also kinds of fungus. The plural of fungus is fungi.

This is the fungus that causes athlete's foot, magnified many times.

foot

Your foot is at the lower end of your leg. It is joined to your leg by the ankle joint. On the end of your foot you have five toes that help you to **balance**. See also **human body**.

fracture

A fracture is a break in a **bone**.

funny bone

Your funny bone is in your **elbow**. It is the end of a bone in your upper arm called the humerus. If you knock your funny bone, you get a tingling pain in your elbow. It is called the funny bone because the name humerus is like the word humorous, which means 'funny'. See also **human body**.

bone pierces skin

bone shatters

clean break

incomplete break (greenstick)

These are some of the ways in which an arm bone may fracture or break.

G

gall bladder

Your gall bladder is a small bag near your **liver**. It stores **bile**. See also **digestion**.

gene

grandmother

Your genes are tiny parts of the **chromosomes** in every **cell** of your body. Genes are made of **DNA**, which controls the way in which you grow. Your genes are a mixture of copies of your parents' genes, and the DNA in them makes you grow to look like your parents.

mother

genetic

Genetic means anything to do with **genes** and inheritance. The study of how parents pass characteristics, such as eye colour, on to their children is called genetics.

daughter

△

In this family, similar features have passed from one generation to the next.

germ

Germs are very tiny living things. They may cause diseases or **infection**. **Bacteria** and **viruses** are germs.

German measles

German measles is another name for **rubella**.

gland

Your glands make and store the different kinds of fluids your body needs to live and grow. Some glands make fluids, like **sweat** and **saliva**, which pass along tubes called ducts to different parts of the body. These glands are called exocrine glands. Other glands, called endocrine glands, make **hormones** which go straight into your blood.

▷

These are some of the main glands (shown in red) in your body.

pituitary

thyroid

ovaries (female)

testes (male)

pancreas

glasses

Glasses are two pieces of special glass or plastic in a frame that people may wear to correct their **vision**. Glasses are also called spectacles.

goose pimples

Goose pimples are the little bumps that come up on your skin when you are cold. See also **shivering**.

graze

A graze is slight damage to the skin caused by rubbing or scraping that makes it bleeds a little. It should be cleaned with an **antiseptic** to remove any dirt. Abrasion is another name for graze.

groin

Your groin is the part of your body where your **abdomen** meets the top of your legs.
See also **human body**.

growing

Growing means not only that your body gets bigger but also that it changes. People usually keep growing until they are about 20 years old. The pituitary **gland**, which is at the base of your brain, produces a **hormone** that makes you grow.

gullet

Your gullet is the tube which carries food to your **stomach** when you swallow.
Oesophagus is another name for gullet.
See also **digestion**.

gums

Your **teeth** grow out of your gums. It is a good idea to brush your gums as well as your teeth to keep them clean and healthy.

gut

Gut is another name for the **intestine**.

New hairs starting to grow from the scalp.

Two hairs, magnified many times, growing out of the scalp.

haemophilia

Haemophilia is a disease of the **blood**. It is passed on to a person from their parents. If someone with haemophilia cuts themselves, the cut does not heal very quickly and they may lose a lot of blood. This is because the blood cannot clot properly.

follicle

hair

Hair grows out of little pits in your **skin** called follicles. It helps to protect your skin. A hair is made of a substance called keratin and contains dead cells. However, the base of each hair is alive, and new cells push older ones upwards to make the hair grow longer.

Hair grows all over your body, except the palms of your hands and the soles of your feet. Most hair grows on your head.

As you grow up extra hair begins to grow under your arms and boys begin to develop a beard on their chin. Also at this time, **pubic** hair starts to grow in your **groin**, near a bone called the pubis. As a person grows older, their hair may turn white and may also fall out.

follicle

Straight hair, wavy hair and curly hair grow from different shaped follicles.

follicle

hand

Your hand is on the end of your arm and is joined to it by the wrist **joint**. On the end of each hand there are four fingers and a thumb. There are many different ways in which you can bend and move your hands and this lets you perform all kinds of actions.
See also **human body**.

hay fever

Hay fever is an **allergy** caused by pollen from grasses or flowers. It makes you sneeze and causes your eyes to get red and watery.

headache

A headache is a pain in your head. It is often caused by being tired and will go away if you rest. But if a headache continues, it is a good idea to see a doctor to find out the cause.

headlice

Headlice are tiny insects that live in **hair** and make the head itch. They move easily from one person to another and so can spread very quickly in schools. The eggs of headlice are called nits. Special shampoos and fine combs help to get rid of headlice and nits.

△ *This is one headlouse magnified many times.*

health

Your health is the state of your body and your mind. To help to keep in good health, you need to eat properly, to take exercise and to keep clean.
See also **food**, **hygiene**.

hearing

Hearing is being able to listen to sounds with your ears.
See also **deafness**.

▽
The left side of your heart receives fresh blood (shown in red) and the right side receives used blood (shown in blue).

heart

Your heart is an organ made of **muscle** which lies inside your chest between your lungs. It is about the same size as your clenched fist, and pumps **blood** around your body. It is divided into four parts called chambers. The two upper chambers are called the *atria* and the two lower ones are called *ventricles*.

The left side of the heart receives blood full of **oxygen** from your lungs and pumps it into your main artery, the aorta. From there it travels all around the body. The right side receives used blood and pumps it back to the lungs. In the lungs the used blood gives up **carbon dioxide** and takes up **oxygen**. See also **circulation**, **pulse**.

aorta

right atrium

valves

right ventricle

left atrium

valves

left ventricle

▽ *The sequence of movements that takes place as your heart beats.*

① *Blood enters atria from lungs and body.*

② *Atria contract forcing blood through valves into ventricles.*

③ *Ventricles contract forcing blood out of heart to lungs and body.*

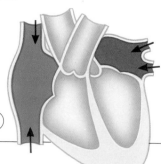
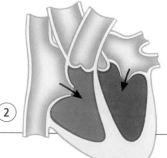
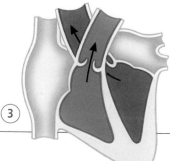

① ② ③

height

Your height is how tall you are. You grow taller as you get older, until you are about 20 years old. Boys and girls grow at different speeds. See also **growing**.

hepatitis

Hepatitis is a disease of the **liver**. A person with hepatitis often has **jaundice** so that their skin turns yellow.

This chart shows the average height of boys and girls at different ages.

heredity

Heredity is the way parents pass on certain characteristics, like the colour of eyes, to their children before birth. The children are said to have inherited these characteristics. See also **chromosome, DNA, gene.**

hernia

A hernia happens when some part of the inside of the body sticks out of the place where it is supposed to be. It happens because the **muscles** holding the part in place have become weak. Hernias sometimes have to be repaired by an **operation**.

pelvis

thigh

hiccups or hiccoughs

Hiccups are sudden gasps of breath that make a 'hic' sound. They are caused by your **diaphragm** tightening suddenly. After each breath, your vocal cords in your **larynx** snap shut and this causes the noise. Hiccups may continue for some time, but usually stop fairly quickly.

hip joint

hip

Your hip is the large joint which joins the top of your leg to your **pelvis**. See also **human body**.

HIV

HIV is the **AIDS** virus. HIV stands for Human Immunodeficiency Virus.

hole in the heart

Sometimes people are born with a hole between the right and left sides of their **heart**. It causes blood that is low in **oxygen** to travel around the body, which can be dangerous. Doctors will sew up a hole in a baby's heart as soon as possible.

hormone

Hormones are chemical substances which help to control the way your body works and grows. They are made in your body by your endocrine **glands**. These glands pass hormones into your **blood**. See also **adrenaline, diabetes.**

Human Body

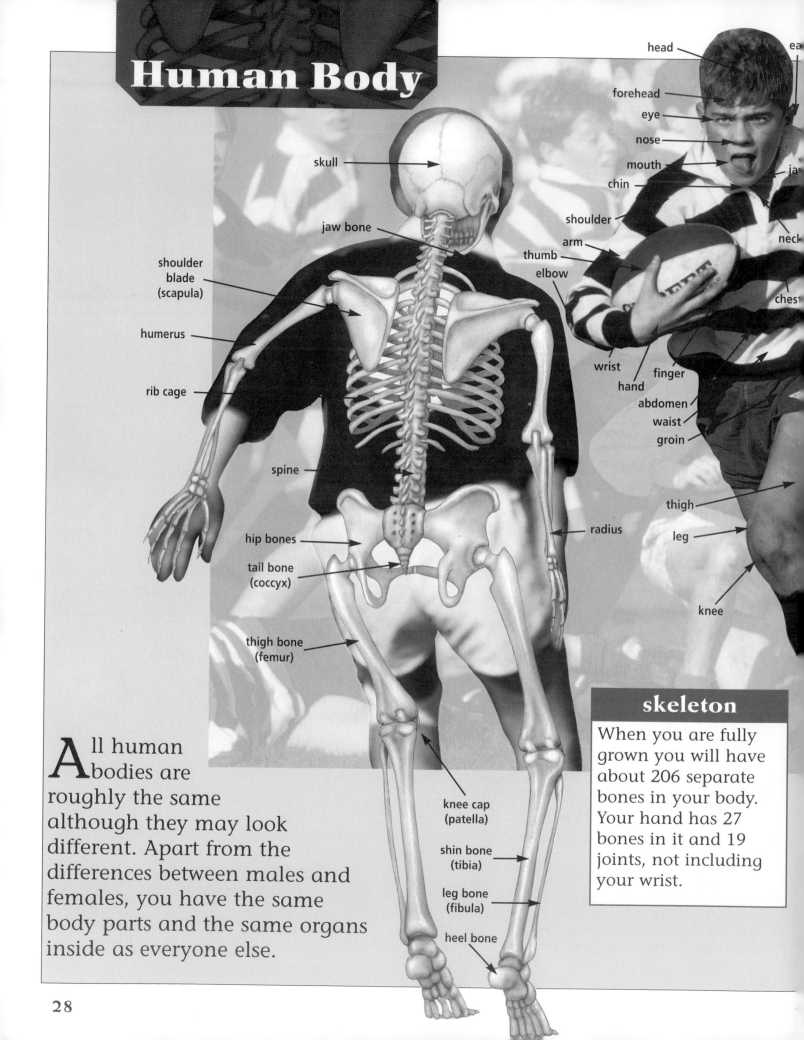

skull

jaw bone

shoulder blade (scapula)

humerus

rib cage

spine

hip bones

tail bone (coccyx)

thigh bone (femur)

knee cap (patella)

shin bone (tibia)

leg bone (fibula)

heel bone

head

forehead

eye

nose

mouth

chin

shoulder

arm

thumb

elbow

wrist

hand

finger

abdomen

waist

groin

thigh

radius

leg

knee

ea

ja

neck

ches

All human bodies are roughly the same although they may look different. Apart from the differences between males and females, you have the same body parts and the same organs inside as everyone else.

skeleton

When you are fully grown you will have about 206 separate bones in your body. Your hand has 27 bones in it and 19 joints, not including your wrist.

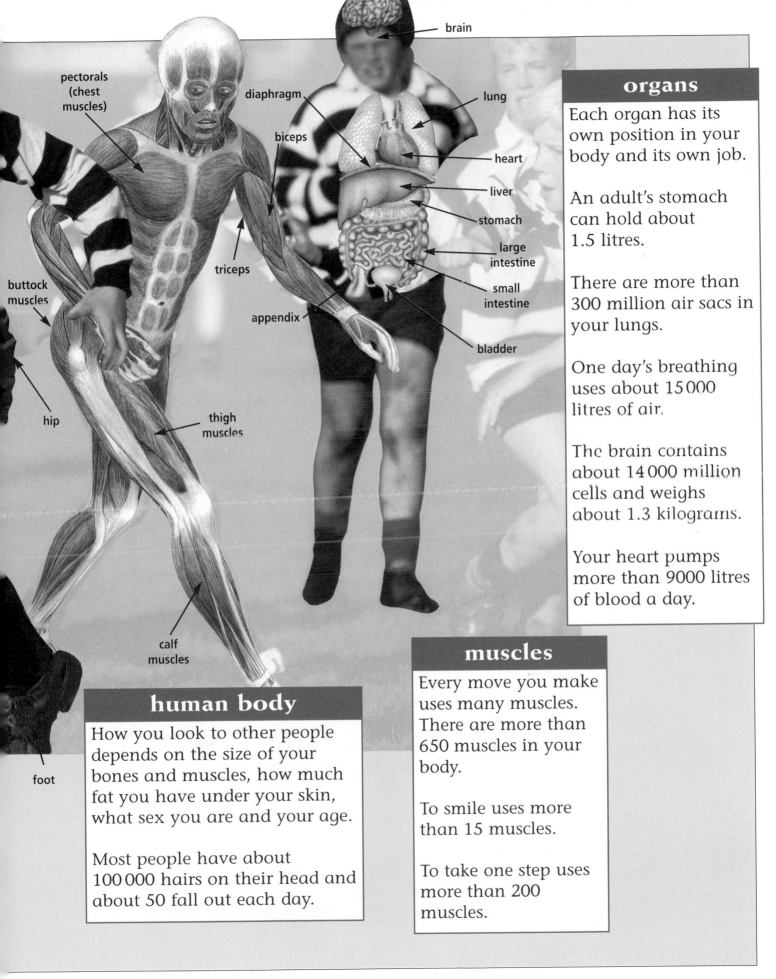

brain

pectorals (chest muscles)

diaphragm

biceps

lung

heart

liver

stomach

large intestine

small intestine

triceps

buttock muscles

appendix

bladder

hip

thigh muscles

calf muscles

foot

organs

Each organ has its own position in your body and its own job.

An adult's stomach can hold about 1.5 litres.

There are more than 300 million air sacs in your lungs.

One day's breathing uses about 15 000 litres of air.

The brain contains about 14 000 million cells and weighs about 1.3 kilograms.

Your heart pumps more than 9000 litres of blood a day.

muscles

Every move you make uses many muscles. There are more than 650 muscles in your body.

To smile uses more than 15 muscles.

To take one step uses more than 200 muscles.

human body

How you look to other people depends on the size of your bones and muscles, how much fat you have under your skin, what sex you are and your age.

Most people have about 100 000 hairs on their head and about 50 fall out each day.

hygiene

Hygiene is keeping your body clean to help it stay healthy. It also means keeping your clothes and surroundings clean so that harmful **germs** cannot grow.

hypnosis

Hypnosis is a way of making a person seem to go to sleep, but in such a way that they hear and obey instructions or suggestions. Sometimes hypnosis is used to help people to stop smoking or to lose weight.

hypodermic

Hypodermic means under the skin. Sometimes a syringe is called a hypodermic because it injects something under the **skin**.
See also **injection**.

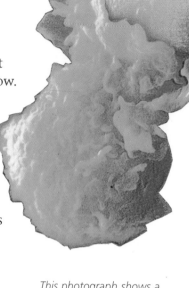

△ This photograph shows a white blood cell destroying a germ in the body.

I

illness

Illness is being sick or having a **disease**. When you have an illness, you feel unwell. A *physical* illness affects the body and a *mental* illness the mind.

immune system

Your immune system is the way in which your body defends itself against **disease**. This is mainly done by your white blood cells, which attack and destroy harmful **bacteria** and **viruses** that enter your body. Immunization helps your immune system to attack certain diseases.
See also **blood**.

immunization

Immunization is a way of helping the body to fight certain **diseases**. This is done by giving people **vaccinations**, usually by **injection**, and sometimes by mouth. The injections provide substances which protect a person against **infection**.

▽ This chart shows the ages at which children are usually immunized against certain diseases.

impetigo

Impetigo is a skin disease which causes small blisters to form. It is easy to catch impetigo from other people. It should be treated by a doctor.

incisor

Your incisors are your big front **teeth**.

indigestion

Indigestion is a pain in the stomach that comes because your body cannot properly digest the food you have eaten.
See also **digestion**.

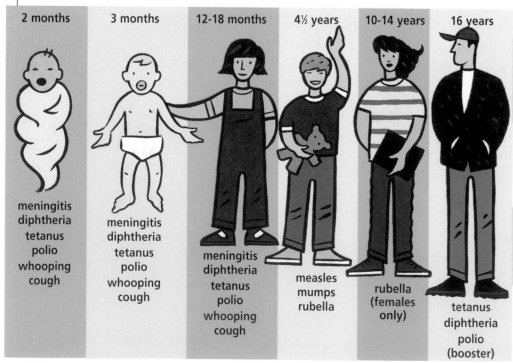

2 months	3 months	12-18 months	4½ years	10-14 years	16 years
meningitis diphtheria tetanus polio whooping cough	meningitis diphtheria tetanus polio whooping cough	meningitis diphtheria tetanus polio whooping cough	measles mumps rubella	rubella (females only)	tetanus diphtheria polio (booster)

infection

An infection is an illness caused by harmful **germs** entering the body. It spreads from one person to another, or from dirty surroundings or food that is not fresh. Something that contains harmful germs is said to be infected.

inflammation

Inflammation is a way in which parts of the body react to disease or injury. A diseased or injured part becomes swollen and painful, and may look red. An insect bite makes your skin become inflamed.

influenza

Influenza is a disease caused by a **virus**. A person with influenza may have a sore throat, a headache, a runny nose and a **fever**. A slight attack of influenza is like a cold, but it can be more serious. It is very easy to catch influenza from other people. Influenza is usually called flu.

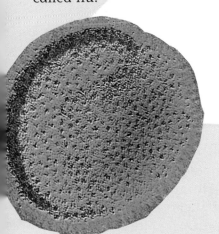

A single influenza virus, magnified many times.

injection

An injection is a way of putting a liquid, such as a **drug**, into your body through your skin. Sometimes the liquid is injected under your skin and this is called a hypodermic injection. A special instrument called a **syringe** is used to give injections.

insomnia

Insomnia is being unable to **sleep**.

insulin

Insulin is a hormone that is made in the **pancreas**. It controls the amount of sugar in your blood.
See also **diabetes**.

intelligence

Intelligence is how well a person is able to think and to understand. A person with high intelligence can learn things and solve problems more easily.

intestine

Your intestine is a long tube coiled up inside your **abdomen**. It is in two parts, which are called the small intestine and the large intestine. Your food passes slowly through the intestine after it has left the **stomach**. See also **digestion, human body**.

small intestine

large intestine

appendix

iris

The iris is the coloured part of your **eye**.

Although your small intestine is narrower than your large one, it is longer and may be up to four metres in length.

itching

Itching is when your skin tickles and you want to scratch it. It can be caused by insect bites, by flakes of dry skin, by an **allergy** or by an **infection**.

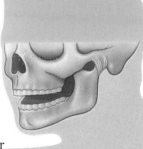

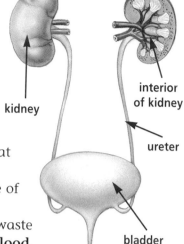

Thin tubes called ureters join kidneys to your bladder.

jaundice

When a person has jaundice, their skin and the whites of their eyes turn yellow. This is because there is something wrong with their **blood** or **liver**. A doctor or nurse has to test the blood of a person with jaundice to find out what is wrong. The treatment will depend on the cause of the jaundice.

jaws

Your jaws are the lower part of your face. They consist of the upper jaw, which is fixed to your **skull**, and the lower jaw, which moves. Your **teeth** are attached to your jawbones. The lower jawbone is connected to the skull by **joints** in front of your ears so that it can move about when you eat and talk. See also **human body.**

The lower jaw is the only part of your skull that moves.

keratin

Your **hair** and **nails** and the outer layer of your **skin** are made of a hard substance called keratin.

kidney

You have two kidneys at the back of your **abdomen** on each side of your backbone. The kidneys get rid of waste and water from your **blood**.

kidney

interior of kidney

ureter

bladder

The waste liquid they make is called urine, which flows from the kidneys and collects in your **bladder.**

Sometimes a person's kidneys do not work properly and they have to use a special machine to clean their blood. This process is called dialysis and the machine is called an artificial kidney. It is possible to live with only one kidney so people sometimes give one of their healthy kidneys to someone else. This is done by a **transplant** operation.

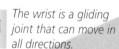

The wrist is a gliding joint that can move in all directions.

gliding joint

The hip is a ball and socket joint that can move in any direction.

ball and socket joint

joint

A joint is the point where two or more **bones** meet. Joints allow you to bend different parts of your body. Your knees and elbows are joints. The bones in a joint are held together by strong bands called ligaments. The ends of the bones are covered with pads of **cartilage** to protect them. A special fluid keeps them 'oiled' so that they move smoothly.

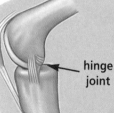

hinge joint

The knee is a hinge joint that can only move up and down.

knee

Your knee is the joint between your upper and lower leg. See also **human body.**

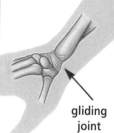

In these illustrations the bones are shown yellow and ligaments green.

labour

Labour is the stage during **birth** when a pregnant woman gradually pushes her baby out of the womb.

laryngitis

Laryngitis is a mild illness that is caused by an inflamed larynx.

larynx

Your larynx is in your throat. It is also called the voice box because your voice comes from it. When you speak, air passes over the vocal cords inside your larynx, and they make a sound. Your lips, tongue, teeth, cheeks and your throat muscles all help you to form these sounds into words.
See also **Adam's apple**, **speech**.

leg

Your leg is made up of the thigh, knee and lower leg. It is joined to the main part of your body by the hip joint and to your foot by the ankle.
See also **human body**.

leukaemia

Leukaemia is a serious disease of the **blood**. It is a kind of **cancer**.

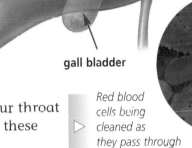

▽ *Your liver plays an important part in your digestive system.*

diaphragm

right lobe

gall bladder

▷ *Red blood cells being cleaned as they pass through the liver.*

left lobe

ligament between the lobes

lice

Lice are tiny insects that may live on parts of the body, causing itching and **inflammation**. Headlice are the most common. A single one is called a louse. See also **parasite**.

liver

Your liver is the largest organ in your body. It is in the top right-hand side of your **abdomen** under your ribs. Your liver takes part in **digestion**, and helps to keep your blood clean. It receives substances that come from your food and drink, and stores some until they are needed and turns others into more useful substances. Some foods and some drinks contain harmful substances, called toxins, and the liver changes these into harmless ones.
See also **bile**.

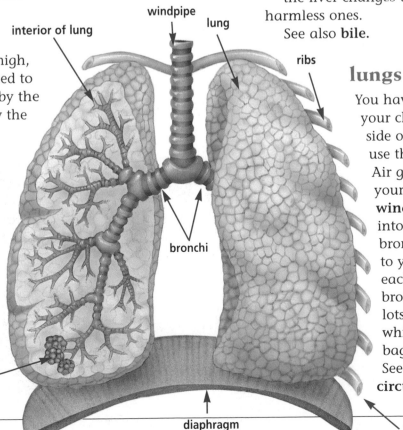

windpipe

lung

interior of lung

bronchi

ribs

lungs

You have two lungs in your chest on either side of your **heart**. You use them to breathe. Air goes in through your nose, down your **windpipe**, and then into two tubes called bronchi which lead to your lungs. Inside each lung, the bronchi branch into lots of little tubes which end in tiny bags called air sacs.
See also **breathing**, **circulation**.

▷ *There are about 300 million air sacs in your lungs.*

air sac

diaphragm

rib

This mosquito, which causes malaria, is sucking blood from a human arm.

M

malaria

Malaria is a disease that is common in tropical places. It can cause headaches, shivering, sweating, and pains in the arms and legs. Malaria is treated with **drugs**, but often returns later. Certain types of mosquitoes carry malaria and spread the disease by biting people. People in tropical countries and visitors take special tablets to protect themselves from malaria. They also sleep under nets at night to avoid being bitten by mosquitoes.

marrow

Marrow is a jelly-like substance inside your **bones**. It makes red and white **blood** cells.

ME

ME is the short name for an illness called myalgic encephalomyelitis. People with ME may feel very tired and weak most of the time. They are often ill for a very long time. Doctors are trying to find a cure for ME but they still do not known what causes it.

measles

Measles is a common disease among children. It is caused by a **virus** and is easily caught from other people. Measles starts rather like a cold, but after a few days red spots appear on the skin. It can be treated with **antibiotics**. Children can be protected against measles by a **vaccination**.

medicine

Medicine is the study of diseases and how to cure them. People study medicine in order to become doctors. Medicine also means the substances, such as **antibiotics** or **drugs**, that doctors give to people to help cure their illnesses.

memory

Your memory is information stored in your **brain**. Long-term memory lasts for many years and helps you recognize your friends and remember where you live. Short-term memory only lasts for a short time, such as remembering the kind of food you ate yesterday. When people get old, they often cannot remember things as well as they did when they were younger. Sometimes an illness or an accident, such as a bang to the head, can cause a person to lose all or part of their memory. This is called **amnesia** and usually the memory comes back.

In this scan, showing the interior of the head, memory is in the part of the brain coloured green.

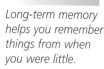

Long-term memory helps you remember things from when you were little.

meningitis

Meningitis is an illness which affects the brain and spinal cord. It is caused by infection from a **bacterium** or a **virus**. The thin membranes called the meninges, which cover the brain and spinal cord, become inflamed. Meningitis can be a very serious disease, causing fever, severe head pains and vomiting. It needs to be treated very quickly with **antibiotics**.

menopause

The menopause is the time when a woman stops producing **eggs** from her **ovaries** and so stops having **periods**. This usually happens between the ages of about 45 and 50. The menopause is sometimes called the change of life.

menstruation

Menstruation is another name for **period**.

mental illness

If someone becomes too upset and unhappy to carry on with their usual life, they are said to have a mental illness. They may behave in an unusual way. Special doctors called psychiatrists look after people with a mental illness.There are different kinds of mental illness. Some are caused by problems in the person's life. Others may be due to disease of the brain or damage to it. People may also be born with a mental illness.

metabolism

Metabolism is all the processes that go on inside your body to keep it alive and to make it grow, heal and repair itself. Your metabolic rate is the speed with which the cells of your body use energy to carry out all these processes. Metabolism is controlled by the **thyroid** gland, which makes the **hormone** thyroxine. Thyroxine controls how your body uses the energy it gets from food.

microbe or micro-organism

This is a tiny living thing that can only be seen when magnified under a microscope. Microbes are present in your body and are all around you. **Bacteria** and **viruses** are microbes.

These microbes live in your intestines.

midwife

A midwife is a person with special training to help a woman who is giving **birth**. The midwife also looks after and examines the woman before the baby is born to make sure that she and the unborn baby keep healthy.

To test your short-term memory, shut the book and see how many of these objects you can remember.

migraine

A migraine is a kind of **headache**. Some people have migraines regularly. Before a migraine starts, the person may seem to see bright flashes of light or wavy lines. The headache is usually very painful and the person may also be sick. Sometimes **drugs** are used in the treatment of migraine.

minerals

Minerals are substances that your body needs to keep healthy and grow. They come from your food. Iron is a mineral needed to form red **blood** cells. Calcium and phosphorus make strong **bones** and **teeth**. Other minerals are needed in small amounts. As long as you eat a healthy **diet**, you should get all the minerals your body needs. See also **food.**

MINERALS	FOUND IN	NEEDED BY
iodine		thyroid gland
iron		blood
calcium		bones
phosphorus		bones
sodium		kidneys
potassium		valves

miscarriage

A miscarriage is when an unborn baby dies and leaves its mother's **womb** in the early part of **pregnancy**. This may happen because there is something wrong with the baby.

molar

Your molars are your big back **teeth** that you use for grinding up your food.

mole

A mole is a small dark mark on the skin, sometimes with a few hairs growing from it. People may have moles when they are born or they may develop later. Most moles are harmless. But if a mole changes in any way, it is a good idea to ask a doctor to look at it.

mouth

Your mouth is the opening in your face that you use for talking, eating and sometimes for breathing. Your mouth is surrounded by your lips and inside are your **gums**, **teeth**, palate and **tongue**. The palate is also called the roof of the mouth. Your tongue helps you to break up your food and also to taste food and drink. You move your tongue, lips and teeth to form words as you speak. The inside of your mouth is kept damp by **saliva**.

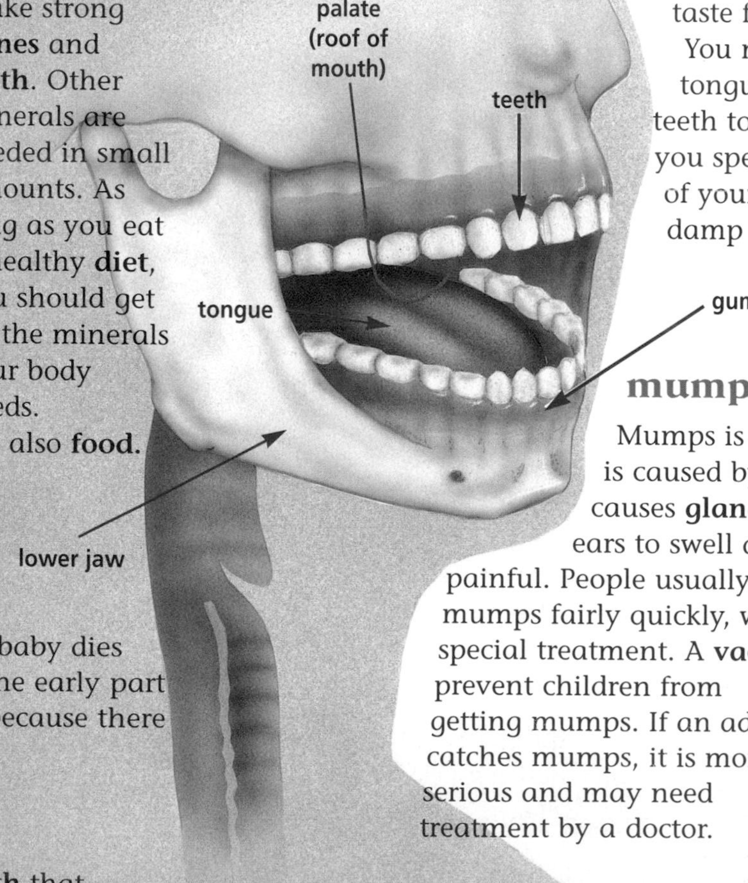

You open and close your mouth by moving your lower jaw.

palate (roof of mouth)

teeth

tongue

gums

lower jaw

mumps

Mumps is a disease that is caused by a **virus**. It causes **glands** around the ears to swell and become painful. People usually recover from mumps fairly quickly, without any special treatment. A **vaccination** can prevent children from getting mumps. If an adult catches mumps, it is more serious and may need treatment by a doctor.

Mumps is caused by this virus. Three are shown here, magnified many times.

Muscles

Muscles are the parts of your body that make you move, so that you can walk or write, for example. Muscles also make parts inside your body work, like your **lungs** and **heart**. You have more than 600 muscles in your body.

Each muscle can contract, or tighten, and get shorter, and then relax to lengthen and go back to its original shape. Muscles are attached to many body parts, and move these parts as they contract and relax. Many of your muscles are joined on to your bones by bands called tendons.

The muscles that move the bones are called voluntary muscles, which means that you control them yourself. The muscles of your **stomach** and **intestine** work by themselves and you cannot control them. These are called involuntary muscles. The muscles of your heart are called cardiac muscles and they are also involuntary.
See also **human body**.

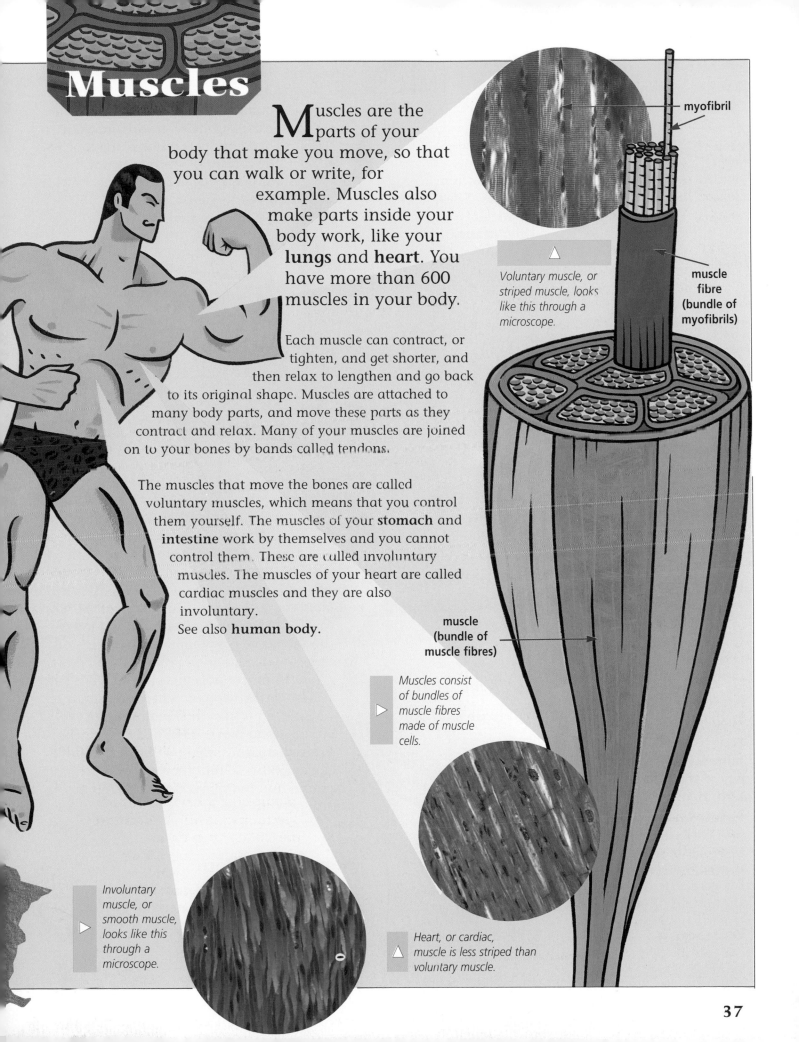

myofibril

Voluntary muscle, or striped muscle, looks like this through a microscope.

muscle fibre (bundle of myofibrils)

muscle (bundle of muscle fibres)

Muscles consist of bundles of muscle fibres made of muscle cells.

Involuntary muscle, or smooth muscle, looks like this through a microscope.

Heart, or cardiac, muscle is less striped than voluntary muscle.

37

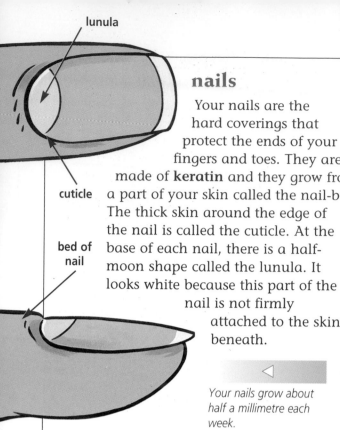

lunula

cuticle

bed of nail

nails

Your nails are the hard coverings that protect the ends of your fingers and toes. They are made of **keratin** and they grow from a part of your skin called the nail-bed. The thick skin around the edge of the nail is called the cuticle. At the base of each nail, there is a half-moon shape called the lunula. It looks white because this part of the nail is not firmly attached to the skin beneath.

◁

Your nails grow about half a millimetre each week.

N

neck

Your neck is between your shoulders and your head. It supports your head and lets it move about. Important blood vessels run through your neck to bring blood to your brain.
See also **human body.**

muscle

nerve carrying message from skin to spinal cord

skin

nausea

Nausea is a feeling of wanting to be sick, or wanting to vomit. It can be caused by eating too much, or by eating bad food, by **travel sickness** and various other things. Once you have been sick, you often feel better. If the feeling of nausea continues, you may need to see a doctor.

navel

Your navel, or tummy button, is the place where you had an **umbilical cord** before you were born. An unborn baby gets nourishment from its mother through this cord. When the baby is born, the doctor or **midwife** cuts the cord. The remaining short stump falls off after a week or so, leaving the navel.
See also **birth.**

△

This shows how a nerve branches and joins with muscle fibres in a muscle.

nerve

A nerve is a bundle of fibres or strands made up of nerve **cells** or neurones. There are nerves in your skin, inside your eyes, ears, mouth and other parts of your body. They enable you to feel, see, hear and taste things. Your nerves, **brain** and **spinal cord** make up the nervous system. Nerves branch off the spinal cord to all parts of the body. Your brain controls the whole of your body through the nervous system. Messages consisting of electric signals flash to the brain along nerves from body parts. The brain then sends messages back to control the body parts so that they work properly. The brain and the spinal cord make up the central nervous system.
See also **reflex action.**

nervous system

Your **brain**, **spinal cord** and all the **nerves** in your body make up your nervous system.
See also **central nervous system.**

nettle rash

Nettle **rash**, or urticaria, is a red, itchy rash caused by stinging nettles. The name is also given to rashes caused by **allergies**. Nettle rash can be treated with special creams or lotions.

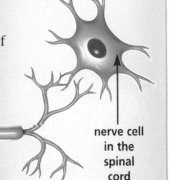

These are nerve cells in the spinal cord.

nerve carrying message from spinal cord to muscle

nipples

Your nipples are the small raised parts on each side of your **chest**. In women they are in the centre of each **breast**. A baby sucks milk through its mother's nipples. See also **breast**.

nerve cell in the spinal cord

Electric signals flash along the nerves and carry messages between the brain, spinal cord and the body.

nit

The eggs that are laid in the hair by **headlice** are called nits.

nose

Your nose is in the middle of your face. You use it to breathe and to smell. The two holes in your nose are called nostrils. As you breathe in air through your nose, it is cleaned and warmed before it goes on its way to your **lungs**. There is a hollow space at the top of the inside of your nose called the nasal cavity. Nerve endings in this space pick up smells and send messages to your **brain**, so that you smell things.
See also **smell**.

nosebleed

Nosebleed is bleeding from one or both nostrils. It can be caused by a knock to the nose.

If a nosebleed does not stop quickly, it can be treated by pinching the nostrils together for about 10 minutes. If your nose is bleeding it is best to sit up, and you should try not to sniff or blow your nose. If the bleeding does not stop, ask an adult for help.

numbness

Numbness is a loss of feeling in a part of the body. It sometimes happens to the fingers or toes when it is very cold. This is because blood does not reach the tips of the body fast enough. The feeling usually comes back quite quickly as the person warms up.

Parts of the body may also go numb if they are pressed on for too long, for example if you lie on your arm in bed. This kind of numbness happens because the nerves get squashed and cannot work properly. As the nerves start to work again you may get a tingly feeling, which is sometimes called '**pins and needles**'.

skull

nerve endings

part of brain sensing smell

nasal cavity

Most of your nose is made of cartilage, not bone.

mouth

nose

O

oesophagus

Oesophagus is another name for **gullet**.
See also **digestion**.

operation

An operation is a way of mending parts of the body. Sometimes organs need to be removed or repaired because they are damaged or diseased. A doctor called a surgeon carries out operations. First the patient is given a **drug** called an anaesthetic, which stops them feeling any pain or makes them unconscious. Then the surgeon can cut open their skin and do the operation. Afterwards the cut edges are sewn up so that the skin heals easily.
See also **unconsciousness**.

organ

An organ is a part of the body which carries out a particular job. Your organs include your heart and lungs, and also parts like your eyes and ears which are organs of sight and hearing.
See also **human body**.

ovary

An ovary is one of a pair of organs in a woman's body that produce egg **cells** or ova. The ovaries are in the lower part of a woman's abdomen.
See also **conception**, **period**, **reproduction**.

oxygen

Oxygen is a gas in the air and humans need it to live. When you breathe air into your **lungs**, some oxygen goes from the air into your blood and travels to the cells in your body.
See also **cell**, **circulation**, **yawning**.

P

paediatrics

Paediatrics is the study and treatment of children's diseases. A doctor who mainly treats children is called a paediatrician.

egg leaves ovary

▷ About every 28 days a woman's ovaries release an egg.

inside of ovary

womb

ovary

vagina

pain

You feel pain when your body hurts somewhere. If you hurt yourself or some part of you hurts inside, pain messages are carried to your brain by your nerves. When an injury heals or you recover from an illness, the pain goes away. Drugs called pain killers help to stop pain.

Pain can help you when you accidentally hurt yourself, for example by touching a hot object. You feel pain and immediately move away from the danger, so that your body is caused no more harm.

pain killer

A drug that helps to stop you feeling pain is called a pain killer.

palate

Palate is another name for the roof of your **mouth**.

pancreas

The pancreas is a **gland** in the abdomen. It makes a **hormone** called **insulin**, which controls the amount of sugar in your blood. The pancreas also makes a juice which helps to digest food in the small intestine.
See also **diabetes, digestion, human body.**

paralysis

Paralysis is not being able to move part or all of the body. It can happen if the brain is damaged by injury, disease or by a **stroke**, so that it cannot send messages along the nerves that make muscles move. Sometimes paralysis is caused by damage to the nerves that lead to or from a part of the body and the brain.

bladder

parasite

A parasite is a tiny living thing that may live on the skin or inside the body. If people do not keep themselves clean, they may get tiny insects called fleas, mites or lice living on them and biting them.
There are also tiny worms that sometimes live inside the body and can cause itching when the person goes to the toilet.
Parasites can be got rid of with special medicines.
See also **headlice, hygiene.**

mite

pelvis

Your pelvis is a strong ring of bone made up of the lower part of your **spine** and your two **hip** bones. It is the place where your legs are attached to the rest of your body. It also helps to protect the organs inside your **abdomen**. A woman's pelvis is usually wider than a man's so that there is room for a baby to grow.
See also **human body.**

testicles

penis

penicillin

Penicillin is a drug called an **antibiotic**. It is used to treat illnesses like **pneumonia**, a disease of the lungs from which people often used to die.

penis

The penis is an organ which only boys and men have. It is at the top of the legs above the **testicles**. When a boy or man goes to the toilet, they pass urine through the tip of their penis. The penis is also used in **reproduction** when sperm pass from the penis into the woman's body.
See also **circumcision.**

period

A period or menstruation is when blood comes from a girl's or woman's **vagina** each month. It is completely normal and natural. When a girl is about 10 to 14 years old, she starts to develop into an adult. Every month, her **ovaries** release an egg **cell**, which travels to the **womb**. If the egg does not join a sperm, the egg and the lining of the womb are passed out through the vagina as blood. Women continue to have periods until they are between the ages of about 45 and 50. The time when they stop is called the **menopause** or the change of life.
See also **puberty, reproduction.**

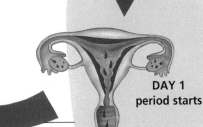

DAY 14-20
egg enters
womb

egg

lining
builds up

If the egg released by an ovary doesn't join with a sperm a woman or girl has a period, which can last up to six days.

egg

DAY 14
egg released

DAY 1
period starts

Some bacteria produce poisons.

perspiration

Perspiration is another name for **sweat**.

phobia

A phobia is being very afraid of something that is not really dangerous. Some people are afraid of the dark, others of insects, spiders, snakes or birds. Although they know that none of these things can hurt them, people with a phobia cannot stop themselves feeling scared.

physiotherapy

Physiotherapy is a treatment that exercises a part of the body that has been injured in some way. The patient is trained to do special exercises by a physiotherapist. These exercises strengthen **muscles** and **joints** to help them to work properly again.

Sometimes the physiotherapist uses massage, which is a special kind of rubbing that can help to loosen muscles and joints.

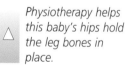

Physiotherapy helps this baby's hips hold the leg bones in place.

pink eye

Pink eye is another name for **conjunctivitis**.

pins and needles

Pins and needles is a tingly feeling you get when your nerves start to work after a part of your body has been numb.
See also **numbness**.

placenta

The placenta is a **tissue** that grows in the **womb** of a pregnant woman. One side of it is attached to the wall of the womb. The other side is joined to the unborn baby by the **umbilical cord**. See also **pregnancy**.

plasma

Plasma is a yellowish liquid that is part of the **blood**. The blood cells and platelets float in plasma.

plastic surgery

Plastic surgery is a treatment for damaged or scarred parts of the body. It covers up or repairs injuries so that the patient looks better. Sometimes doctors can cover scars by using skin from somewhere else on the body. This is called a skin graft. Other kinds of injury can be repaired by using plastic materials.

pneumonia

Pneumonia is a disease of one or both of the **lungs**. The lungs become sore and parts of them fill with liquid making breathing difficult. If both lungs are affected the disease is called double pneumonia. **Antibiotics**, such as **penicillin**, are used to treat pneumonia.

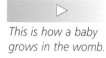

This is how a baby grows in the womb.

1st month

3rd month

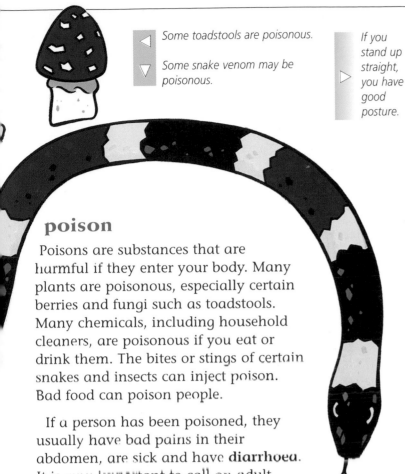

Some toadstools are poisonous.

Some snake venom may be poisonous.

If you stand up straight, you have good posture.

posture

Your posture means the position of your body when you are standing, sitting or moving. If you let your body droop and slump, it may put a strain on your muscles and make you ache.
See also **muscles**.

poison

Poisons are substances that are harmful if they enter your body. Many plants are poisonous, especially certain berries and fungi such as toadstools. Many chemicals, including household cleaners, are poisonous if you eat or drink them. The bites or stings of certain snakes and insects can inject poison. Bad food can poison people.

If a person has been poisoned, they usually have bad pains in their abdomen, are sick and have **diarrhoea**. It is very important to call an adult immediately if you think someone may have been poisoned. Some things are so poisonous that a person can die if they do not get treatment straight away.

pregnancy

Pregnancy is the time from when a baby starts to grow inside its mother until it is born. This usually takes about nine months from **conception**. A woman with an unborn baby is pregnant, and she stops having **periods** during pregnancy.

As the baby develops in the mother's **womb**, her **abdomen** starts to swell. Inside the womb, the baby is joined to the **placenta** by the **umbilical cord**, and the placenta itself is attached to the wall of the womb. Food and oxygen pass from the mother to the baby through the placenta and umbilical cord, which also take away waste substances. When the baby has grown enough, it is ready to be born.
See also **birth, conception, reproduction.**

polio

Polio is the short name for a disease called poliomyelitis, or infantile paralysis. It is caused by a **virus** which affects the nerves that control muscles. It causes **fever**, a sore throat and a headache. Sometimes it can be very serious and may cause **paralysis**. Polio is now a rare disease because most children are immunized against it.
See also **immunization.**

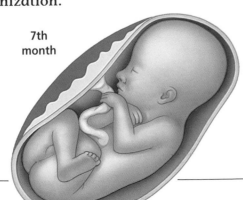

5th month

7th month

9th month

umbilical cord

placenta

womb

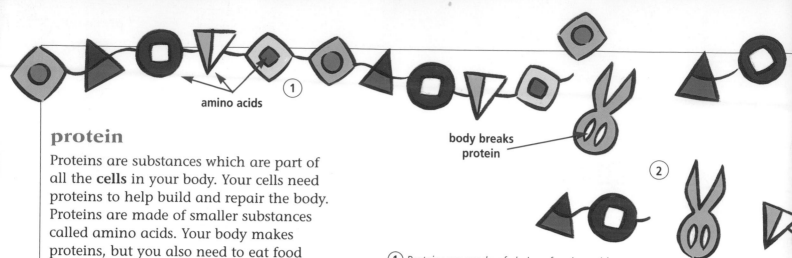

amino acids

body breaks
protein

protein

Proteins are substances which are part of all the **cells** in your body. Your cells need proteins to help build and repair the body. Proteins are made of smaller substances called amino acids. Your body makes proteins, but you also need to eat food containing proteins so that you have all the amino acids you need. The proteins are changed into amino acids when they are digested, and then the amino acids are used by your cells to build new proteins.

① Proteins are made of chains of amino acids.
② Your body breaks the chains of proteins into amino acids.
③ Your body builds new proteins by linking up the amino acids of the proteins you have eaten.

psychiatry

Psychiatry is the treatment of **mental illness**. Doctors who treat mental illness are called psychiatrists.

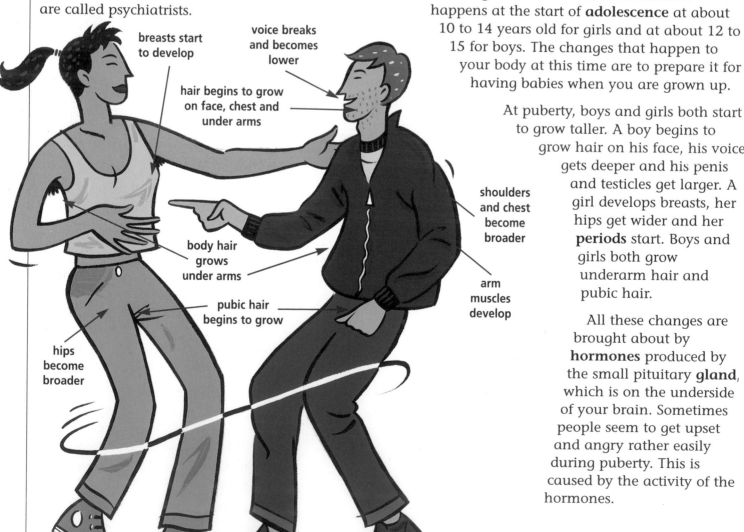

breasts start
to develop

voice breaks
and becomes
lower

hair begins to grow
on face, chest and
under arms

body hair
grows
under arms

shoulders
and chest
become
broader

arm
muscles
develop

pubic hair
begins to grow

hips
become
broader

puberty

Puberty is the time in your life when you begin to change from a child into an adult. It happens at the start of **adolescence** at about 10 to 14 years old for girls and at about 12 to 15 for boys. The changes that happen to your body at this time are to prepare it for having babies when you are grown up.

At puberty, boys and girls both start to grow taller. A boy begins to grow hair on his face, his voice gets deeper and his penis and testicles get larger. A girl develops breasts, her hips get wider and her **periods** start. Boys and girls both grow underarm hair and pubic hair.

All these changes are brought about by **hormones** produced by the small pituitary **gland**, which is on the underside of your brain. Sometimes people seem to get upset and angry rather easily during puberty. This is caused by the activity of the hormones.

pubic

Pubic refers to the part of your body in your **groin**. The word comes from the name of a bone called the pubis, which is at the front of your **pelvis**.

pulse

Your pulse is the throbbing you can feel if you put your fingers on your wrist or your neck. It is caused by your **heart** pumping blood around your body. A ten-year-old child's heart pumps, or beats, about 90 times every minute. A baby's heart beats much faster than this and an adult's heart is a little slower. When you are excited or if you are exercising, your pulse goes faster. The speed of your pulse might change if you are unwell, which is why the doctor may measure it.

pupil

Your pupil is the round black hole in the centre of the iris of your **eye**. Light enters your eye through the pupil, which gets wider in dim light and smaller in bright light. The size of the pupil is changed by muscles in the iris in order to control the amount of light entering the eye.

pus

Pus is a thick, yellowish liquid that collects around an infected part of your body. It is made up of dead blood cells and germs. Spots may contain pus. Pus can be cleaned away with **antiseptic**.

body builds new protein

Q

quadruplets

Quadruplets, or quads, are four babies who grow inside their mother's womb at the same time and are born one after the other. The birth of quadruplets is very unusual.

See also **reproduction, twins**.

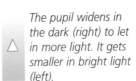

Only one set of quadruplets is born for every six of twins.

The pupil widens in the dark (right) to let in more light. It gets smaller in bright light (left).

quarantine

Quarantine is a time in which you have to keep away from most other people. This may be necessary when you have an infectious disease or if you have been in contact with a person who has an infectious disease.

You could have caught the disease, but it may be some time before you start to feel ill. This time is called the incubation period, and during it you can pass on the infection to other people.

If you have not developed the disease at the end of the incubation period, you can come out of quarantine and mix with other people again. The incubation period is different for each disease.

R

rabies

Rabies is a very serious disease caused by the bite of an animal infected with the rabies **virus**. Anyone who is bitten by an animal that might have rabies needs treatment at once. The disease can be treated with a vaccine but this is not always successful. People visiting or living in countries where there are animals with rabies can be vaccinated against the disease.
See also **vaccination**.

radiation

Radiation is invisible rays of energy. Some forms of radiation are used in the treatment of certain diseases, such as **cancer**. Too much radiation is very dangerous because the rays will damage or kill healthy living cells.
See also **X-ray**.

radiotherapy

Radiotherapy is the treatment of a disease with radiation, such as X-rays.
See also **radiation**.

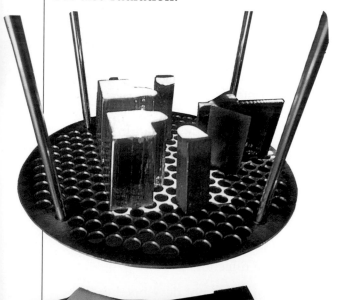

rash

You have a rash when you have a large number of small spots on your skin. Rashes often itch, and can be soothed with special creams. A rash may be caused by an **allergy**, or it may be a **symptom** of a disease like chicken pox or measles.

reaction

A reaction is the way in which you behave when something happens to you. For example, some people may laugh when they are a little scared. This is called a nervous reaction.

Reaction also means the way in which your body behaves when it is in contact with certain substances. The rash or sneezing caused by an **allergy** is an allergic reaction. Some **drugs** may cause your body to have an allergic reaction, and the doctor will tell you not to take them. If someone is cured of an illness by taking medicine, they are said to have a good reaction to the medicine.

rectum

Your rectum is at the lower end of your large **intestine**. When waste is ready to leave your body, it goes into the rectum. From there, it passes through the **anus** and leaves your body as **faeces** when you go to the toilet.
See also **digestion**.

reflex action

A reflex action, or a reflex, is an action your body makes automatically. Your reflexes help to protect you from danger. If something burns your fingers, you pull them away immediately, without thinking about it. This is a reflex action. It happens because your **nerves** flash a message directly to your muscles, telling them to move your hand away.
See also **pain**.

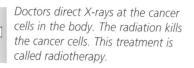

Doctors direct X-rays at the cancer cells in the body. The radiation kills the cancer cells. This treatment is called radiotherapy.

reproduction

Reproduction is the way in which babies are made by men and women. An **egg** cell produced by a woman must meet a **sperm** cell produced by a man before a baby can start to grow. It is during sexual intercourse that a sperm passes into a woman's body and joins with the egg. This joining is called **conception**, or fertilization. Once an egg is fertilized it starts to grow into a baby in its mother's womb. See also **pregnancy, sex.**

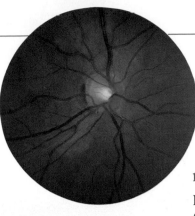

This is a close-up photograph of the retina of the eye.

retina

Your retina is the layer of cells at the back of your **eye**. These cells are sensitive to light.

rheumatism

Rheumatism is a general name for any painful **inflammation** or stiffness of the **joints**.

rib

A rib is one of the thin, curved bones you can feel in each side of your chest. You have twelve pairs of ribs, which are joined to your spine at the back. Ten pairs are joined on to your breastbone in front, and two pairs come only half way round and are called 'floating' ribs. See also **breathing.**

mouth

lung

ribs

diaphragm

Oxygen passes from the air through your lungs to your body.

respiration

Respiration means the way in which the cells in your body use the oxygen from the air and substances from food to produce energy and carbon dioxide.

Respiration is also another word for **breathing**, when you take oxygen into your body and let carbon dioxide out.

rubella

Rubella is a disease caused by a **virus**. It is also called **German measles**. It causes a pink **rash** on the face, neck and body. It is not a serious disease for children to catch, but it can be harmful to the unborn baby of a pregnant woman if she catches it. Most girls are given vaccinations against rubella when they are about 13 years old, so that they will not catch the disease when they are adult and might be pregnant.

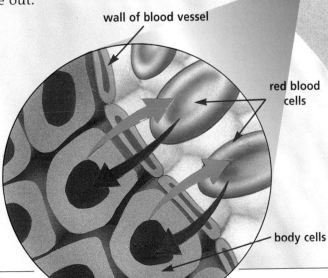

wall of blood vessel

red blood cells

body cells

Red blood cells throughout the body deliver oxygen to body cells and take up carbon dioxide from the body.

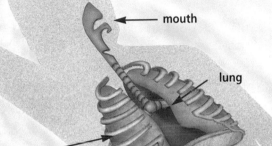

sacrum

Your sacrum is a part of your **spine**. It is a triangular bone, made up of five vertebrae joined together. It is at the back of your **pelvis**.
See also **spine**.

A drop of saliva seen through a microscope.

saliva

Saliva is the watery liquid in your mouth. It is also called spit. Some saliva is made all the time by **glands** in your mouth, but most is made when you are eating food. The saliva mixes with your food to make it easy to swallow. An **enzyme** in saliva helps the digestion of some kinds of foods. Saliva also helps to keep your mouth clean.

scab

A scab is the hard crust that forms over a **wound**. The scab protects the wound from infection, and falls off when the wound is healed.

scar

A scar is the mark left on the skin after a wound has healed. If a scar is very noticeable, it may be treated by **plastic surgery**.

scarlet fever

Scarlet fever is a disease causing a sore throat and a red rash. It mostly affects children and is caused by **bacteria** infecting the **tonsils**. Scarlet fever is usually treated with **penicillin**.

senility

Senility is a loss of ability that some old people suffer. They may be very muddled and forgetful, and this is called 'becoming senile'. It may happen because the person's brain is not getting enough oxygen to keep it healthy.

senses

Your senses let you know what is happening in the world around you. You have five main senses: **sight**, **hearing**, **touch**, **smell** and **taste**. Your sense organs are your eyes, ears, skin, nose and mouth. Your ability to **balance** is also called your sense of balance.

Your senses give you pleasure, and also warn you of danger. You need your senses to do things, such as run, dance and eat. Blind and deaf people lack one sense, but they are often able to use other senses to make up for this loss.

In this picture, the most sensitive parts of your body, such as your fingers and lips, are drawn larger than the less sensitive parts.

serum

Serum is a clear yellow liquid that separates from **blood** as it clots. It is like **plasma** but does not contain the substances that make blood clot. Serum is used in the treatment of some diseases, and it can be injected to protect people against certain diseases or poisons.

sex

Your sex is whether you are male or female. The sex of a person is decided by their **chromosomes**.

Sex is also a short way of saying sexual intercourse. This is when a man puts his penis into a woman's vagina and passes sperm into her body.
See also **conception, reproduction.**

shivering

Shivering is a shaking movement that happens if you get cold. It actually helps your body to keep warm. Shivering is a **reflex action** caused by tightening of certain muscles. At the same time, other tiny muscles make the hairs on your skin stand up to trap warm air between them. The little bumps this raises on your skin are called **goose pimples.**

shock

Shock is a sudden lowering of blood pressure which stops your blood flowing properly around your body. It can be caused by losing a lot of blood, by being in an accident or by having a serious upset or bad fright. A person suffering from severe shock goes very pale and cold, and their **pulse** speeds up and becomes weak.

shoulder

Your shoulder joins your arm to the rest of your body.
See also **human body.**

son

mother

father

daughter

Female eggs have X chromosomes and male sperms X and Y. At conception, these combine to give a male (XY) or a female (XX).

sight

Sight is being able to see everything around you. You use your **eyes** to see and the measure of how well you do this is called **vision**. If you have poor vision, you cannot see objects clearly. With short sight, you can only see nearby objects well. With long sight, only distant objects are clear. You may need to wear glasses or contact lenses to correct poor vision.

sinus

Your sinuses are air-filled spaces inside your **skull**. They are in your forehead, behind your nose and in your cheeks. They warm the air you breathe in. When you speak, the air inside your sinuses vibrates, or shakes. This affects the tone of your voice and helps it to sound clearer.

skeleton

Your skeleton is made up of all the **bones** in your body. It is the framework that holds your body together, keeping it up and giving it shape. Many of your **muscles** are fixed to your skeleton, and parts of your skeleton surround and protect the **organs** inside your body.
See also **human body.**

A baby has more than 300 bones in its skeleton, but some of them join together as it gets older. An adult's skeleton contains about 206 separate bones.

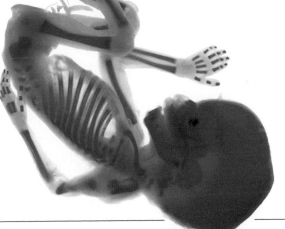

Skin

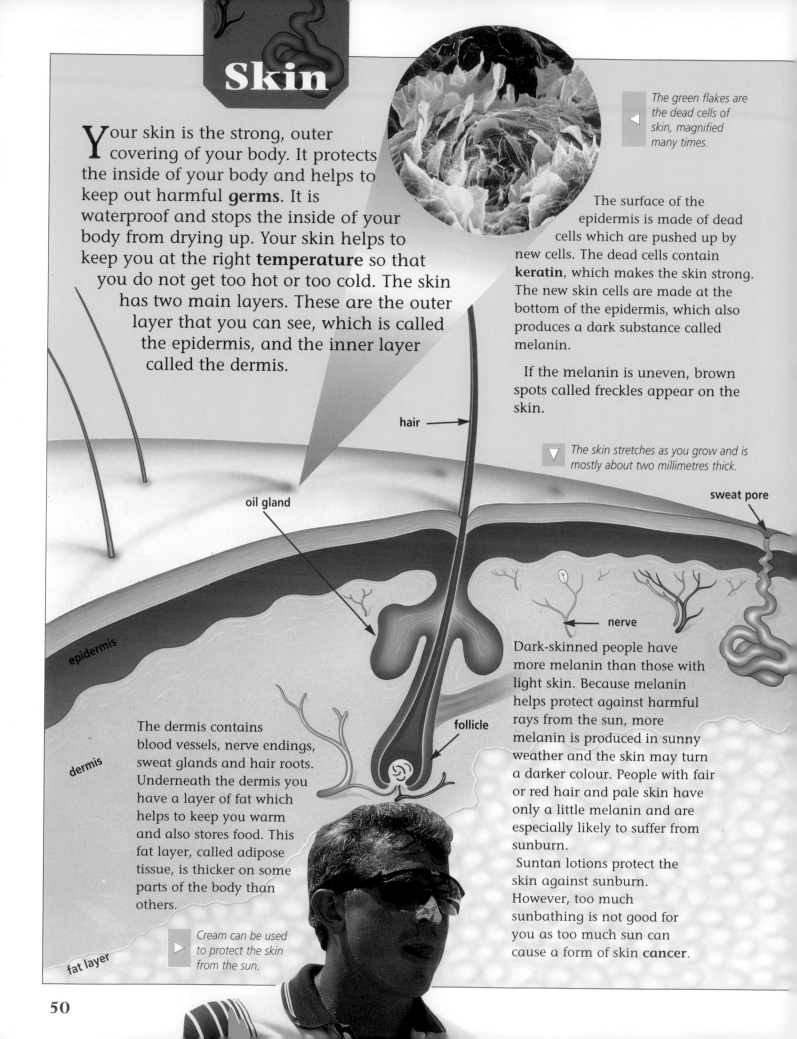

Your skin is the strong, outer covering of your body. It protects the inside of your body and helps to keep out harmful **germs**. It is waterproof and stops the inside of your body from drying up. Your skin helps to keep you at the right **temperature** so that you do not get too hot or too cold. The skin has two main layers. These are the outer layer that you can see, which is called the epidermis, and the inner layer called the dermis.

The green flakes are the dead cells of skin, magnified many times.

The surface of the epidermis is made of dead cells which are pushed up by new cells. The dead cells contain **keratin**, which makes the skin strong. The new skin cells are made at the bottom of the epidermis, which also produces a dark substance called melanin.

If the melanin is uneven, brown spots called freckles appear on the skin.

The skin stretches as you grow and is mostly about two millimetres thick.

hair

oil gland

sweat pore

nerve

epidermis

follicle

dermis

The dermis contains blood vessels, nerve endings, sweat glands and hair roots. Underneath the dermis you have a layer of fat which helps to keep you warm and also stores food. This fat layer, called adipose tissue, is thicker on some parts of the body than others.

Dark-skinned people have more melanin than those with light skin. Because melanin helps protect against harmful rays from the sun, more melanin is produced in sunny weather and the skin may turn a darker colour. People with fair or red hair and pale skin have only a little melanin and are especially likely to suffer from sunburn.
Suntan lotions protect the skin against sunburn. However, too much sunbathing is not good for you as too much sun can cause a form of skin **cancer**.

Cream can be used to protect the skin from the sun.

fat layer

skull

Your skull is the hard, bony box that gives your head its shape. It is made up of 29 bones, some of which are joined, or fused, together. The main part of the skull, called the cranium or 'brain box', contains your brain and protects it from damage.

joined bones

The rest of the skull consists of 14 bones that shape your face and jaw, three tiny bones in each ear and one bone at the base of your tongue.
See also **human body.**

sleep

Sleep is a time when your body is at rest. It is not the same as being unconscious because you can easily be woken up. When you are asleep, your brain works far more slowly than when you are awake. Your heart and your breathing slow down and your muscles relax.

There are different kinds of sleep. In deep sleep, your brain is very relaxed and your body is still. During deep sleep, growth hormones are released into your blood. Growing children need these hormones, and this is one of the reasons why children sleep more than adults.

Sometimes people find it difficult to sleep even when they feel tired. This is called insomnia. It may happen because the person is worried about something, or because they are ill. A doctor will treat someone suffering from insomnia.
See also **unconsciousness.**

In a baby's skull, the cranium is made up of eight separate bones that join together when the baby becomes a young child. The places where these bones join are called fixed joints or sutures.

smell

Your sense of smell lets you pick up different scents from the air around you. When you sniff something, the air with the scent goes up your **nose** into the space called the nasal cavity. Inside this cavity there are cells with tiny hairs on them. The hairs are covered in sticky substance that absorbs the scent. The cells then send messages through nerves to your brain, and you detect the scent.

Your sense of smell is closely linked to your sense of **taste.** If your nose is blocked, so that you cannot smell very well, you will probably find that you cannot taste properly either.

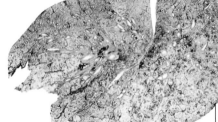

Look at the difference between the inside of a non-smoker's lung (bottom) and a smoker's lung (top).

smoking

When people smoke cigarettes, cigars or a pipe, they breathe in tobacco smoke. Most doctors believe that smoking is very bad for you. Tobacco smoke contains several harmful substances, including tar and nicotine. Tar makes the linings of the air passages leading to the **lungs** produce extra mucus. Smokers often cough a lot as they try to clear the mucus away. This can lead to **bronchitis** and a very serious lung disease called emphysema. Nicotine can affect the nervous system and can also cause heart disease. **Cancer** of the lungs, caused mainly by smoking, kills many people each year.

sneezing

You sneeze in order to clear quickly something that is tickling your nose. Sneezing is a **reflex action**. You breathe in deeply and then the air is blown out of your nose and mouth with great force. Sneezing helps to spread diseases because tiny drops of mucus containing **bacteria** and **viruses** fly out with the air. This is why you should sneeze into a handkerchief.

◁ *When you sneeze, the air travels at more than 160 kilometres per hour.*

snoring

Snoring is the snorting or grunting noises some people make when they are sleeping. If you sleep with your mouth open, the air you breathe in can make your throat muscles vibrate. Also, a flap of skin at the back of the throat, called the soft palate, may rattle as you breathe out. Some people snore so loudly they wake themselves up!

sore

Sore has two meanings. If you feel sore, it means that a part of your body is painful when you touch it. Soreness can be caused by injuries, or by spots or boils. A burn makes your skin sore.

A sore is another word for a painful place on your body, such as an **abscess** or an **ulcer**.

▽ *Your larynx protects the vocal cords in your throat.*

larynx

windpipe

spastic

Spastic describes a kind of paralysis in which some of the muscles go stiff. **Cerebral palsy** causes a kind of spastic paralysis.

speech

The sounds you make when you talk are called speech. You use your **larynx**, or voice box, to speak. Your larynx has two bands of **cartilage** stretched across it. These are your **vocal cords**. As you speak, you breathe out and air passes over the vocal cords making them vibrate to produce sounds. Muscles in the larynx change the shape of the vocal cords so that the sounds can be high or low. The sounds are then formed into speech by moving your mouth, tongue and lips. See also **sinus**.

sperm

Sperm are the male sex cells. They are produced in a man's testicles. During sexual intercourse, millions of tiny sperm cells travel through the man's penis into the woman's body. When one joins with the woman's egg cell, a baby starts to grow. See also **reproduction, sex**.

◁ *Sperms' tails enable them to swim and join the egg in a woman's body.*

spina bifida

Spina bifida is a defect of the spine that some people are born with. It can cause **paralysis** in the legs or damage to the brain. It is possible for doctors to test unborn babies to see if they have spina bifida.

spinal cord

Your spinal cord is the thick bundle of **nerves** that runs from the base of your **brain** down to the bottom of your back. With your brain, it makes up your central nervous system. Nerves branch out from the spinal cord to every part of your body. The spinal cord is the main pathway for nerve messages to and from your brain. It is protected by the vertebrae in your **spine**.

The vertebrae that make up your spine each have a hole through which the spinal cord passes.

spinal cord · vertebra · disc between vertebrae · nerve

spine

Your spine is the long chain of bones that forms your backbone. These bones, called vertebrae, stretch from your neck to the bottom of your back. Between each pair of vertebrae there is a cushion of **cartilage** called a disc, and the spine is held together by **muscles** and **ligaments**.

sacrum

Near the bottom of an adult's spine five vertebrae have fused together to form the sacrum, which is the back of the pelvis. Below the sacrum is the coccyx. or tail-bone, which, in an adult, is made up of four fused vertebrae.

coccyx

spit

Spit is another name for **saliva**.

spleen

Your spleen is in the left side of your body between your **stomach** and your **diaphragm**. It does several things. The spleen produces some of the **blood** for unborn babies. It also makes white blood cells that help to defend your body against harmful **germs**. Your spleen also removes old or damaged cells and other waste from your blood.

spots

A spot is a small, inflamed swelling on your skin. You get spots when the tiny openings, or pores, in your skin, get blocked. Sebum, an oil made by glands in your skin, builds up behind the blockage and a spot forms. You may get spots during **puberty** because the skin makes more sebum at this time. Spots are also called zits. It is best not to squeeze spots because this may let in harmful germs and cause infection.
See also **acne**.

sprain

A sprain is an injury to a **joint**, such as your wrist or ankle. The joint becomes twisted and the ligaments that hold the bones in place get stretched or torn. A bandage helps to support the joint while the ligaments heal. The ankle is sprained more often than any other joint.

squint

A person has a squint if both **eyes** do not point in the same direction. It can happen if the muscles that move the eyeball are not working properly. A squint can sometimes be corrected by exercises to improve the action of the muscles. Special glasses can also help, and sometimes people have a small operation to alter the muscles.

stammering

Stammering, or stuttering, is a speech disorder. People who stammer have difficulty in speaking smoothly because they repeat the beginning of a word, or the whole word, over and over again. This happens because muscles in the **larynx** are not working properly. People who stammer can be helped by special treatment called speech therapy.

stethoscope

A doctor uses a stethoscope to listen to noises inside your body, especially in your **heart** and **lungs**. The stethoscope helps the doctor to decide what is wrong with you when you are ill.

earpiece

stitch

A stitch is a pain you sometimes feel on the left side of your body just below your **ribs**. It can happen when you are running over rough ground or running when your stomach is full. No one really knows exactly what causes stitch, and it goes away when you stop running.

gullet

stomach wall

small intestine

stomach lining

The lining of your stomach is covered with many tiny pits. Glands in these pits help you to digest your food.

The stethoscope has two earpieces and a piece that rests against the patient's body.

stomach

Your stomach is in the upper part of your **abdomen**. It is sometimes called your tummy. Your stomach is an important part of your digestive system.

Sometimes you may get a pain in your abdomen that is called stomach ache. It may be caused by **indigestion**, by **food poisoning** or by an illness. Stomach ache usually goes away quite quickly, but if it doesn't, it is a good idea to see a doctor. See also **digestion**, **human body**.

stroke

A stroke is damage caused to the brain because it does not get a proper blood supply. A stroke often happens suddenly. It may be the result of blocked or burst blood vessels. People usually recover well from a mild stroke. Some strokes are more serious, and can affect a person's speech and movement. A very severe stroke can cause a **coma** and death.

stye

A stye is a painful, red swelling on the eyelid. It is caused by infection around the root of an eyelash. A stye is treated by bathing the swelling with clean, hot water, or sometimes by treating it with **antibiotics**.

sunburn

Sunburn is burning, soreness and blisters on the **skin** caused by sunshine. Sunburn can be treated with soothing lotions and creams.

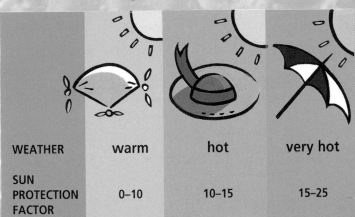

WEATHER	warm	hot	very hot
SUN PROTECTION FACTOR	0–10	10–15	15–25

△ *Sun creams have factor numbers to show the protection they give. This table shows factors needed for different types of weather.*

surgeon

A surgeon is a doctor who is trained to carry out operations.

surgery

Surgery has two meanings. The room in which a doctor sees his patients is often called a surgery.

The branch of medicine that treats diseases and repairs injuries by carrying out operations is called surgery.

▷

On a hot day, you may lose as much as two litres of sweat. Sweat also helps to get rid of waste from your body.

sweat

Sweat is the salty liquid that forms on your **skin** when you are hot. Perspiration is another word for sweat. **Glands** in your skin make sweat to help you cool down. Sweat comes out through tiny holes called pores. As the sweat dries on your skin, it draws warmth from your body and makes you feel cooler.

If you do not wash sweat off your body, for example after taking exercise, it may smell unpleasant as it grows stale. This smell is caused by bacteria and is usually called b.o., or body odour.

◁

Sweat droplets on the surface of the skin magnified many times.

symptom

A symptom is something unusual that happens to your body. It can be something you can actually see, like a rash or a swelling, or it could be a pain or a fever. An illness or an injury often has its own kind of symptoms. A doctor asks patients about their symptoms when deciding what is wrong with them.

syringe

A syringe is an instrument with a tube and plunger at one end and a sharp, hollow needle at the other. It is used to take blood or give drugs.
See also **hypodermic**, **injection**.

▷ *The markings on a syringe show how much liquid it contains.*

taste

Taste is the sense which enables you to tell the flavours of your food and drink. After your food and drink have been mixed with **saliva** in your mouth, the taste of them is picked up by tiny groups of cells called taste buds.

The taste buds are in the sides of the little lumps you can feel on your **tongue**. **Nerves** in these taste buds send messages about the taste to your brain. Your sense of **smell** also helps you to notice flavours.

This map of your tongue shows where the four tastes are detected;
✚ *shows bitter taste*
/ *shows sour taste*
■ *shows salty taste*
❙ *shows sweet taste.*

These are taste buds on the tongue, magnified many times.

TB

TB is short for tuberculosis. It is a serious disease caused by **bacteria**, which first attack the **lungs** but may spread to other parts of the body.

TB used to be very common, but better hygiene and new discoveries about prevention and treatment of the disease mean that now it is fairly rare. TB is treated with **antibiotics** and people can be vaccinated against it.
See also **vaccination**.

A child has 20 milk teeth which are replaced by 32 adult, or permanent, teeth.

tears

Tears are the drops of salty liquid which sometimes run from your eyes. This liquid is produced all the time by **glands** above your eyes. Its main use is to keep the front of each eye clean, and free from dust and harmful germs. The liquid contains a natural **antibiotic** that kills bacteria. When you blink, you spread the liquid over your eye. When you cry, extra liquid is produced and it overflows and runs down your face.

teeth

The teeth in your mouth are used to bite and chop your food before you swallow it. They are made of a substance called dentine covered by a layer of hard enamel. Inside the dentine is a soft part called the pulp cavity, containing blood vessels and nerves. These nerves send messages to your brain if your teeth hurt. The part of a tooth that sticks up above your gum is called the crown. Below your gum, the root of the tooth is fixed into your jawbone by a kind of cement.

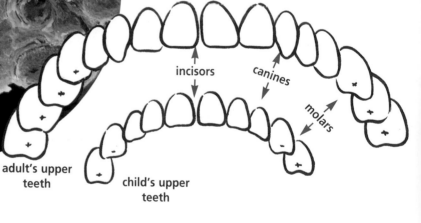

incisors   canines

molars

adult's upper teeth

child's upper teeth

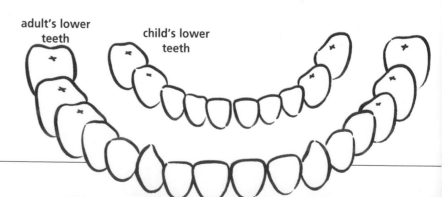

adult's lower teeth

child's lower teeth

Your teeth are so important that you need to look after them properly. Bits of sugary food stuck between them can produce an acid that attacks the enamel, allowing germs to get inside the tooth.

Food and bacteria form a thin layer called plaque on your teeth. If too much plaque builds up, it decays your teeth, making holes in them. **Tooth decay** causes painful **toothache** and needs treatment by a dentist. Dental treatment does not usually hurt.

temperature

Your temperature is a measure of how hot or cold your body is. It is measured by a thermometer. The temperature inside your body should be about 36.5°C (degrees Celsius) or in Fahrenheit 98.2°, but it can be slightly higher or lower than this. However, if you are ill, your temperature may go up or down by a few degrees. This is why a doctor takes your temperature when you are feeling unwell. See also **fever**.

mercury thermometer

tendon

Your tendons are the very strong bands that connect your **muscles** to your **bones**. As your muscles tighten and shorten, the tendons pull on your bones and make them move.

test-tube baby

A test-tube baby is the name given to a baby that grows from an **egg** that has been taken out of the mother's body before **conception** has taken place. The egg and **sperm** are put together by doctors in a laboratory and then put back inside the mother, so that the baby can grow in her **womb**.

digital thermometer

Your temperature can be measured from your mouth or armpit with a mercury or digital thermometer, and from your forehead with a liquid crystal thermometer.

testicle

A testicle, or testis, is one of two sex glands under the **penis** in boys and men. The testicles are egg-shaped and are in a pouch of skin called the scrotum. The testicles produce **sperm** and a male **hormone** called testosterone. During puberty, testosterone controls the growth of body and facial hair in boys.

These are germs that cause tetanus, magnified many times.

tetanus

Tetanus is a very serious disease of the **nervous system**. It is caused by certain germs that live in the soil and in animal **faeces**. The germs get into the body through cuts or scratches and cause the **muscles** to jerk and stiffen. It is sometimes called lockjaw because a person with tetanus might have difficulty in opening their mouth. Fortunately, tetanus is very rare because most people are regularly vaccinated against it. If someone is hurt and there is thought to be a risk of tetanus, they will be given the vaccine straight away. See also **vaccination**.

thermometer

A thermometer is an instrument for measuring **temperature**.

liquid crystal thermometer

thigh

Your thigh is the upper part of your leg. See also **human body**.

thinking

Thinking is using the information stored in your **brain**. When you think, your brain is sorting the information out and deciding what to do with it. By thinking you are able to understand things, to work problems out and to make decisions.
See also **intelligence, memory**.

thorax

Thorax is another name for **chest**.

▽ *The air you breathe and the food you eat both pass down your throat.*

thumb

Your thumb is a part of your hand. It bends towards your fingers so that you can easily hold things.
See also **human body**.

thyroid

Your thyroid is a **gland** in your neck. It is around the front and sides of your windpipe. It produces a **hormone** called thyroxine, which controls how your body uses the energy it gets from food.
See also **metabolism**.

tiredness

A feeling of tiredness means that you need to sleep or to have a rest. If someone still feels tired after a good sleep, or they feel tired most of the time, it could be a **symptom** of an illness and it is a good idea to see a doctor.

bone

tonsils

pharynx

gullet

larynx

tooth

tongue

lower jaw bone

tissues

Your tissues are groups of **cells** that form the different kinds of **organs** in your body. For example, your nerve tissue is made of nerve cells. Other body tissues include **bone**, **skin**, **blood** and **muscle**.

▷ *The cells of ① bone, ② skin, ③ blood and ④ muscles are shown here.*

throat

Your throat is the front part of your neck. The pharynx, which is the back wall of your nose and mouth, extends down your throat. Your **larynx** is in your throat, and also in the openings to your **gullet** and your **windpipe**.

windpipe

toe

You have five toes joined to each foot.

tongue

Your tongue is joined to the back and bottom of your mouth. It is mostly made of muscle, which is why you can move it about so easily.

As well as letting you **taste** your food and drink, your tongue also helps to mash the food up and pushes it down your gullet when you swallow. When you talk, your tongue moves about to help to form the sounds you want to make.

tonsillitis

Tonsillitis is an illness caused by **infection** of the tonsils.

tonsils

Your tonsils are two small lumps right at the back of your mouth, at the entrance to your **throat**. They are there to protect your throat from germs. If your tonsils get infected by **bacteria**, they may swell up, making it difficult to breathe. This is called tonsillitis. Usually they go back to normal when the infection is treated with **antibiotics**. Sometimes a doctor decides that the tonsils should be removed by an operation.

▷

Successful transplant operations have been carried out with these organs.

decayed area
pulp
nerve
root
bone

△

A hole in a tooth caused by tooth decay may reach the part of the tooth where there are nerves, causing toothache.

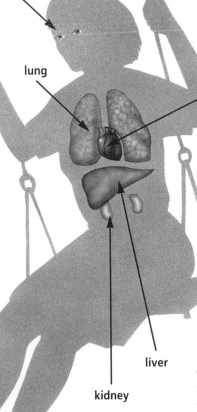

part of eye
lung
liver
kidney

tooth decay

Tooth decay is a disease caused by damage to the enamel covering of your **teeth**. As a result of this damage, holes form in the enamel so that harmful **bacteria** can get inside.

toothache

Toothache is pain in a tooth or in several **teeth**. If you have pain in your teeth it usually means that something is wrong and you should see a dentist.

touch

Touch is the sense which lets you feel things when they come into contact with your skin. You can do this because **nerves** in your skin send messages to your **brain** about the object you are touching. You have millions of these sensory nerves and they tell the difference between hot and cold, hard and soft, and light and heavy touch.

trachea

Trachea is another name for **windpipe**.

heart

transplant

A transplant is an operation in which a healthy **organ** is removed from one person, called the donor, and put into the body of another person, called the recipient.

It is done because the recipient has something wrong with their own organ. Some organs, such as the heart, lungs, eyes or liver, may be taken from the body of someone who has just died. An organ, such as a kidney, may be transplanted from a living person as people can live with only one kidney.

Before a transplant operation, the doctors check to make sure that both donor and recipient have a similar type of **tissue**. If their tissues were very different, the recipient's body would reject the organ and it would not work properly.

travel sickness

Travel sickness, or motion sickness, is a feeling of **nausea** some people get when they are travelling in cars, planes, on ships or on rides at fairgrounds. It happens because the movement disturbs the **balance** organ in each ear and the brain gets confusing messages about what is happening to the body. Children often grow out of travel sickness, but there are also drugs which help to prevent it.

triplets

Triplets are three babies who grow inside their mother's **womb** at the same time and are born one after the other. See also **twins**.

tuberculosis

Tuberculosis is the full name of the illness that is usually called **TB**.

tumour

A tumour is a swelling in or on the body. It happens when a group of cells start to increase in number rapidly, and form a growth. A benign tumour is harmless. But a malignant tumour is a sign of **cancer** and has to be removed or treated.

◁

In this scan of the inside of the brain the part coloured yellow is a tumour.

twins

Twins are two babies that grow inside their mother's womb at the same time. There are two kinds of twins: identical twins, who are very alike, and non-identical or fraternal twins, who are different.

Sometimes, when the mother's egg is fertilized at **conception**, the cells separate into two parts. These two parts grow into two identical babies. Each baby has the same **chromosomes**, so identical twins are always the same sex. They often look so much alike that it is quite difficult to tell which twin is which. Non-identical, or fraternal, twins develop when the mother's **ovaries** release two eggs at the same time. Each egg gets fertilized by a separate sperm cell and two babies develop. Because these twins have different chromosomes they can be of different sex and may not be very alike.

▽ *Guess if these twins are identical or non-identical.*

typhoid

Typhoid, or typhoid fever, is a serious disease caused by **bacteria** in food and drinking water. People with typhoid **germs** do not always develop the disease themselves, but they can easily pass it on to others, especially if they serve food to them. This is one of the reasons why people should always wash their hands very carefully after going to the toilet.

This disease is uncommon in countries where there is good public hygiene and where people are vaccinated against it. Typhoid is treated with **antibiotics**.

vein

artery

umbilical cord

An unborn baby receives nourishment and oxygen through the umbilical cord.

placenta

ulcer

An ulcer is a painful open sore on the skin or inside the body. People often get ulcers in their mouths, especially if they have a rough edge on one of their teeth. These ulcers usually heal by themselves. Another common kind of ulcer is one that develops in the **stomach** because the digestive system is not working properly. A person suffering from this type of ulcer often has to eat a special **diet**, and might need to have an **operation**. People who worry a lot tend to suffer from stomach ulcers.
See also **digestion**.

ultrasound

Ultrasound is sound that is so high you cannot hear it. Ultrasound machines can be used by doctors to make pictures of the inside of a person's body. Parts of the body reflect, or bounce back, the ultrasound waves so that the shape of a part is shown on the screen.

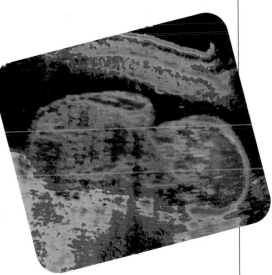

This is an ultrasound picture of an unborn baby seen inside the mother's womb.

umbilical cord

The umbilical cord is a rope-like bundle of blood vessels and nerves that connects an unborn baby to the **placenta** and so to its mother.
See also **navel**.

unconsciousness

Unconsciousness seems like a deep sleep from which a person cannot easily be woken. People may become unconscious as the result of a bang to the head, because they have fainted, or as the result of a fit. If someone becomes unconscious, you should get an adult to help straight away.
See also **coma, fainting, epilepsy**.

vaccination

You are given a vaccination either by an injection, through a small scratch on your skin or by mouth. The medicine in the vaccination is called a vaccine. It prevents you from catching a particular disease.

A vaccine is made from a weak form of the germs that cause the disease. When the vaccine is put in your body you do not develop the disease. Instead, your body makes substances called **antibodies**, which fight the disease if you come into contact with it. For some diseases, this protection wears off after a while and further vaccinations, called boosters, have to be given. Most children in the developed world are given vaccinations against **polio**, **measles**, **rubella** and **TB**.
See also **immunization**.

A vaccination against a disease protects you from the disease.

vagina

The vagina is a passage in the body that only girls and women have. It connects the **womb** and the outside of the body. The opening of the vagina is between the legs. When a baby is being born, the vagina stretches wide open to let the baby out. See also **birth**, **sex**.

varicose veins

Varicose veins are veins that are very swollen and painful and which sometimes need to be operated on.

vein

A vein is any one of the blood vessels in your body that carry used blood back to your heart. See also **circulation**.

This diagram shows how a valve in the vein ① opens; ② closes to prevent blood flowing backwards; ③ becomes weak so blood collects; ④ closes permanently because of the collected blood; and ⑤ swells causing varicose veins.

verruca

Verruca is a name for a kind of **wart**, usually on the sole of the foot.

vertebra

A vertebra is one of the bones that make up your **spine**. The plural of vertebra is vertebrae.

This virus, magnified many times, causes warts.

virus

A virus is a **germ** that causes disease. Viruses are much smaller than **bacteria**. They cannot be killed by **antibiotics**, but it is possible to be vaccinated against many diseases that are caused by viruses. See also **vaccination**.

vision

Vision is the measure of how good your **sight** is.

vitamin

Vitamins are substances that are necessary to keep you healthy. Your body cannot make vitamins, so you have to eat food that contains them. In order to work properly, your body needs small amounts of about fifteen different vitamins. These are usually in the food you eat. But if you do not get the vitamins you need, you suffer from vitamin deficiency and things may go wrong with your body.

vocal cords

Your vocal cords are part of your **larynx**. When you speak, air passes over the vocal cords and they make a sound.
See also **speech**.

voice

Your voice is the sound made by your vocal cords. See also **larynx**.

The vocal cords open (top) and closed (bottom), seen looking down the throat.

vomiting

You vomit when food rushes up from your stomach and out through your mouth. Vomiting is also called being sick. It is a **reflex action** caused by the muscles in your abdomen tightening up and forcing the contents of your stomach back up. Vomiting may be caused by eating too much, but it can also be a symptom of an illness such as **food poisoning**.

People sometimes vomit when they are suffering from **travel sickness**. If someone keeps vomiting, they should be treated by a doctor.

wart

A wart is a small, hard swelling on the skin. Warts are caused by a **virus**. They are usually not painful and often disappear without any treatment. A wart on the sole of your foot, called a plantar wart or verruca, can be painful and may have to be removed.

This virus, magnified many times, causes warts on the hands and the soles of the feet.

weight

Your weight is how heavy you are. It depends mainly on your age, height and sex. Your weight can also depend on your **diet**, or on how much and what you eat. Weight can also be affected by your **glands** and **hormones**. If a person finds it very difficult to remain at the right weight, even with a proper diet, they may need to see a doctor.

whooping cough

Whooping cough is a very infectious disease of children. It is spread by **bacteria** that spray out when an infected person coughs. It is called whooping cough because, after coughing, the person draws in their breath quickly making a 'whoop' sound. Whooping cough is a serious disease, and most children are vaccinated against it.
See also **vaccination**.

windpipe

Your windpipe, or trachea, is the tube that takes air from your mouth down your throat to your lungs. At the top of the windpipe there is a flap called the **epiglottis**. This closes when you swallow anything, so that it does not go down the 'wrong way' and make you choke.
See also **choking**.

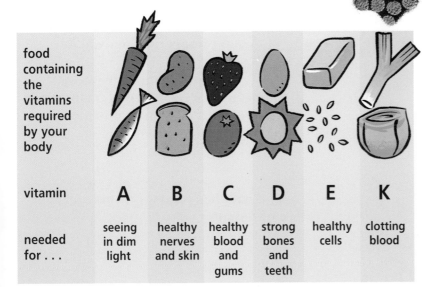

food containing the vitamins required by your body						
vitamin	A	B	C	D	E	K
needed for . . .	seeing in dim light	healthy nerves and skin	healthy blood and gums	strong bones and teeth	healthy cells	clotting blood

wisdom tooth

A wisdom tooth is a molar tooth at the back of the mouth which a person does not usually get until they are over 17 years old.
See also **teeth**.

womb

The womb, or uterus, is an **organ** that only girls and women have. It is in the **pelvis** and is connected to the **vagina** by a narrow tube called the cervix. It is in the womb that a fertilized egg grows into a baby.
See also **birth**, **reproduction**.

worms

Certain kinds of worms are **parasites** that live inside the human body. The more common ones include threadworms, roundworms and tapeworms. They may cause illness. Some kinds cause a lot of itching around the anus.

Worms can be caught from eating some kinds of meat and fish that have not been prepared and cooked properly. Good **hygiene** helps to stop worms, especially washing your hands before eating and after going to the toilet. Worms can be cleared out of the body by special **drugs**.

wound

A wound is damage to the **skin**, which can be a serious cut or a slight **graze**. Your skin is able to repair itself, but doctors sometimes stitch the edges of a deep cut together to help it to heal.

(1) When you cut your skin, it bleeds and (2) your blood forms a clot around the cut. (3) A mesh made of a substance from blood is formed over the cut. This mesh traps the blood cells and forms a scab, (4) which stops bacteria from getting into the cut. (5) The scab drops off when new skin has grown over the cut, and the wound will have healed.

X-ray

X-rays are invisible rays. They are used to take photographs of the inside of your body. The X-rays pass easily through the soft parts of the body but less easily through the **bones** so that the bones show up. The photograph itself is sometimes referred to as an X-ray. X-rays are also used in the treatment of some diseases, especially **cancer**.

This worm is a parasite found in water in some hot countries. It can cause a disease called bilharzia.

An X-ray of a wrist watch inside a person's stomach.

yawning

Yawning is a **reflex action** that makes you open your mouth wide, breathe in, and then breathe out slowly. You do this when you have not been breathing deeply enough to take in as much **oxygen** as you need. This happens when you are tired, or bored or when you are in a stuffy room and need fresh air. Yawning helps you to take in extra oxygen that helps you to stay awake.

zit

A zit is another word for a **spot**.
See also **acne**.